YOU DON'

Arthur Michele received his MD degree from New York Medical College in 1935. He holds a Master's Degree in Science from the State University of Iowa and completed six years of residency training in Orthopaedic Surgery. Dr Michele is Professor and Chairman of the Department of Orthopaedic Surgery at the New York Medical College, and Director of Orthopaedic Surgery at the following hospitals: Flower-Fifth Avenue Hospital, New York City; Metropolitan Hospital, New York City; Byrd S. Coler Hospital, New York City; US Public Health Service Hospital, New York City; and US Public Health Service Hospital, Staten Island. He is also the Honorary Police Surgeon of New York City. He has published numerous articles in many medical journals as well as a breakthrough medical book entitled *The Iliopsoas*.

YOU DON'T HAVE TO ACHE

A Guide to Orthotherapy

ARTHUR A. MICHELE
MD, MS (Orth Surg), FACS

Illustrations by Rosemary Torre

UNABRIDGED
PAN BOOKS LTD : LONDON

First published in Great Britain 1972 by
Souvenir Press Ltd
This edition published 1974 by
Pan Books Ltd
33 Tothill Street, London SW1

ISBN 0 330 23884 1

Printed in Great Britain by
Richard Clay (The Chaucer Press) Ltd, Bungay, Suffolk

This book is dedicated to the young mothers of today who refuse to accept discomfort and awkwardness in their children as normal. Because of them we will have a generation of children who are orthophysically fit.

Acknowledgements

Many thanks to my medical students whose constant questioning has been the main motivation behind my work.

I am grateful to Mr Arne Nicolaysen, who has worked with me for twenty years and is truly the first orthotherapist.

Contents

Foreword

This book is designed to help the millions of men, women, and children who are plagued with aches and pains that do not seem to be helped by ordinary medical care. It is written for the general public, but it will also be of interest to medical students, physicians, nurses and physiotherapists, as well as to everyone in the field of physical education.

The book is the result of my many years of research on muscle imbalance, its causes and treatment, and of the almost miraculous results I have seen when it is corrected. The treatments and results have been reported in articles in medical journals and to physicians at medical meetings, so that the techniques are coming into common use throughout the world. More and more physicians are learning how to recognize the symptoms of muscle imbalance and how to treat them with these new techniques, and as they see their patients' lives dramatically changing, the enthusiasm is spreading.

But there is still an extraordinary gap in medical education concerning the common everyday aches and pains of the musculo-skeletal system. One doctor, complaining of the problem at an American Medical Association meeting, said, 'Muscles and bones make up sixty per cent of the body's organs, but only about one per cent of a medical student's time is spent in learning about them. Then,' he said, 'suddenly we find ourselves in practice faced by patients, about eighty per cent of whom complain to us of some pain related to some part of the musculo-skeletal system.'

The facts about muscle imbalance and the results from the new techniques of treatment are so astounding that, at first glance, one is sceptical. I was even sceptical myself, in the beginning.

I've been in orthopaedic practice for thirty-five years. My training as a doctor, and in years of research, was always to be objective and never to jump over-enthusiastically to false conclusions. From the first day of medical school, a prospective physician is trained to question and evaluate everything. Today I am professor and chairman of the Department of Orthopedic Surgery at New York Medical College and also teach at Flower and Fifth Avenue Hospital and Metropolitan Hospital Center in New York, and I am still questioning. But my long experience has proved that the orthotherapy exercises in this book can bring about what seem to be near-miracles.

Other doctors tend to question and be sceptical, too. But when they have objectively listened to the logic of how aches and pains can be caused by imbalance within the major muscle complexes and they then try the appropriate exercises with one or two patients and see the dramatic results that come about with this simple treatment, they become the most enthusiastic boosters imaginable.

So many people are used to living with their miseries, so used to accepting them as part of their lives, that they don't even think of going to their doctors to ask for help. They say that arthritic pains and back pains are part of growing old, that bow-legs, bunions and sore feet have always run in the family, that children always have growing pains and bad posture. They simply accept as part of their nature the fact that they may be awkward in sports.

But these familiar expressions of resignation are *simply not medical truth*. For adults or children the philosophy that nature will correct their muscle problems or that they have to live with their aches and pains is not only false, but harmful. For it does nothing more than delay the time for treatment so that the condition becomes more and more difficult to correct.

It is my wish that this book will bring a message of hope to the one out of every three people who has muscle problems, will explain muscle imbalance and what it can mean to him and to his family – and will tell him how to recognize it and *treat* it.

Arthur A. Michele, MD, FACS

CHAPTER ONE

You Don't Have to Hurt: What Orthotherapy Is All About

The big stories in medicine that hit the front pages are about cancer and heart disease and stroke. These are tagged as the major problems, the killers.

But the minor problems of daily pain and misery are the ones that are killing our happiness, that are keeping millions of people from leading what should be full and happy lives. These so-called minor problems are taking a back seat in publicity and research, but in everyday life they are in the driver's seat. They are tremendously important, affecting the lives of millions of us, robbing us of pleasure and keeping us from our most productive work.

You've heard it many times. 'We've all got our aches and pains.' Stoic. Accepting.

But we don't have to accept them. The aches and pains and misery aren't necessary. Many people are suffering when they do not have to.

They are needlessly suffering with backache, painful hips, aching feet, splitting headaches, stiff shoulders and neck, cramping pains in their muscles, dislocations, sciatica, weak knees and ankles. They feel awkward in sports, dancing and other activities – even walking.

Their pain can be dull but nagging and constant, a depressing and grim overtone to everything they do. It can cause them to be nervous, tense and irritable, snapping at those they love and dragging through the hours of the day. Or their pain can be so bad that it interferes with their work. It can be so sharp and severe that it sends the sufferer to bed in tears.

I can remember when I was a young doctor and I first saw people suffering these daily pains, I thought, 'People shouldn't *have* to hurt, not on a daily basis.' People do get used to living with their aches and pains. But why should they have to go around constantly with aching feet and hurting back? What a difference it would make if they could get rid of them! It would be like being born again, having a whole new, active life open up.

I can remember walking through the orthopaedic wards at the hospital where patients were in casts and slings, their beds rigged with pulleys and weights like ancient torture racks. All these devices were intended to straighten bones or stretch muscles to eliminate or reduce severe pains or crippling deformities. The children's service was especially poignant. As you walked into the ward you would see twenty-five beds lined up along each wall. In each bed a little girl or boy would be lying face up with his or her small body in a traction harness. Overhead there was a mirror so the child could see visitors and the nurses walking by. Fifty beautiful faces peered at you from the mirrors over the beds. But the eyes were so terribly sad. Most of these children had to be in traction for six months or more to straighten out abnormal curves and crookedness in their spines. Yet, in the majority of cases the abnormalities and crookedness returned as soon as the child became ambulant again.

Early in my career I became aware of the way in which posture could indicate how various muscle groups were functioning, and in turn how muscle function was directly related to overall health and physical fitness. And I became aware, too, that poor muscle function could be altered, and that this alteration depended in turn on the attachment of muscles to joints. It was the point of attachment – the origin and insertion of the muscle – that was significant. In an ancient story, Archimedes is quoted as saying 'Give me a place to stand on and I will move the earth.' The bone that is stationary, that does not move, is the 'place' on which a muscle 'stands', the place that determines how the muscle will make something in the human body move.

The body is built on basic engineering principles. The skeleton is a series of moving parts with bones as the levers and muscles to pull them and make them perform. If there is something wrong with the attachment of a muscle, or if a muscle is too short or too rigid, stresses and strains on our joints are the result. This leads to improper muscle functioning, which produces bad posture, which in turn is the cause of all the resultant aches and pains.

The aching back, sore feet, hip pain, neck and shoulder pains, awkwardness – these are the effects of bad posture. But bad posture is *more* than how you stand. It is the indication of how all of your muscles work.

And – most important of all – *bad posture is not a bad habit.* Contrary to everything you may have been told before, *bad posture is a disorder of the musculo-skeletal system* and the disorder *can be corrected by proper treatment.*

Orthotherapy is the application of specially designed exercises that use the principles of posture and body engineering to bring about the proper functioning of the muscles and the joints. Orthotherapy treats the disorder – *musculo-skeletal imbalance* – of which painful joints and bad posture are merely symptoms.

You May Be Hurting Yourself with Improper Exercise. If you exercise regularly, you may think you are among the most physically fit people in the world. But you may actually be doing exercises that are wrong for you. And you may be doing serious harm to yourself! For if you don't understand the principles of body mechanics you may unknowingly do exercises that could cause irreparable damage to your body.

Two out of three individuals may engage in any sport or exercise activity and excel. However, our current nationwide physical-fitness programmes are strictly geared to this normal group, often to the detriment of the physically unfit, the other one-third of the country.

For these dangerous exercises are those commonly given in most schools, sports, and exercise classes! They are taught in the belief that they are of benefit to everyone. But the fact is that three out of every ten people should not do them!

'Touch your toes and don't bend your knees!' This command goes out in schools, in the army, in women's reducing classes, and in college football warm-up sessions. Some of the exercisers do it with ease, some puff and grunt through it. Others – shamefaced and embarrassed – simply can't make it, and their pride and feelings are hurt. Others are even less lucky – they end up with their bodies hurt, too. The touch-the-toes bit is part of a great exercise error in this country. It is fine for people who have perfectly balanced muscles. But for those with muscle imbalance it can be a disaster because of the tremendous strain it places on the muscles and joints of the lower spine and the hips.

Right at this moment your child could have muscle imbalance that you are unaware of, and his body could be being damaged by his school exercise programme.

Dr James Nicholas, the physician for the New York Jets football team, stressed the dangers of wrong exercise at a recent American Medical Association meeting. He spoke about the two groups of people – the loose-muscled and the tight-muscled – and warned doctors at the meeting of the many people who should never try to touch their toes because they are just not built for it. He cited many injuries caused in school and recreation programmes and called for a national study of the long-lasting injuries produced. Jim Nicholas uses many of the tests described in Chapter III to determine who are the loose-muscled and tight-muscled individuals on the Jets team, and he has the players do many of the same stretching exercises described in Chapter IV.

Another compulsive callisthenic that people with muscle imbalance shouldn't do is press-ups. A lot of children are winning medals for doing press-ups ten, twenty, forty, even fifty times. But by the time they are sixteen or seventeen the group with musculo-skeletal imbalance may become muscle monsters who have extremely developed arms and shoulders with no coordination or leg agility. They were exercising the wrong group of muscles. What might start as a mild contracture of the muscle may eventually become really tight and soon the muscle-boy has to bend forward like an ape because his legs turn out when he walks and he starts developing hip and knee problems to compensate.

If you have a pathological malposture, a weak back or other signs of muscle imbalance, most likely you are what I call a thirty per-cent-er. You should not only avoid certain exercises but even everyday household tasks that put a strain on the back and hip muscles.

For example, here are some things that you should not be doing:

Do not touch your toes standing with your legs straight.
Do not do press-ups or pull-ups.
Do not do sit-ups unless you use a curling motion with your feet free and your knees lightly flexed.
Do not lie face down while raising the arms and legs and arching the back.
Do not lie flat on your back and raise both legs while keeping your knees straight.
Do not climb ropes or ladders, or work on parallel bars.
Do not lift weights.
Do not work with pulleys or steel springs to increase arm and chest muscles.
Do not jump hurdles.
Do not wrestle.

One Out of Three Needs Help. Thirty per-cent of the population – both adults and children – have something wrong with their muscles. This third of the population has aches and pains, or problems with awkwardness or walking, or problems in their sex lives. One out of three people have a muscle imbalance upsetting their lives in one way or another. This muscle imbalance, if recognized, can be treated to bring new hope to these people.

With the instructions in this book you will be able to test yourself and other members of your family for the symptoms of muscle imbalance so you know whether you are among the 'thirty per-cent-ers', as I call them, who need medical help. These tests are simple and are easily done.

If you find you are suffering from a muscle imbalance, you will also find simple methods for correcting your problem.

The Most Important Muscle Syndrome. The iliopsoas, one of the major muscle complexes of the body, is the key to most cases of muscle imbalance. In the diagram below the iliopsoas is striped with fine lines. It goes from the spine through the abdomen and over the brim of the pelvis to the inner part of the upper thigh. Its normal function is involved with the entire working of the back, the hips and the pelvic

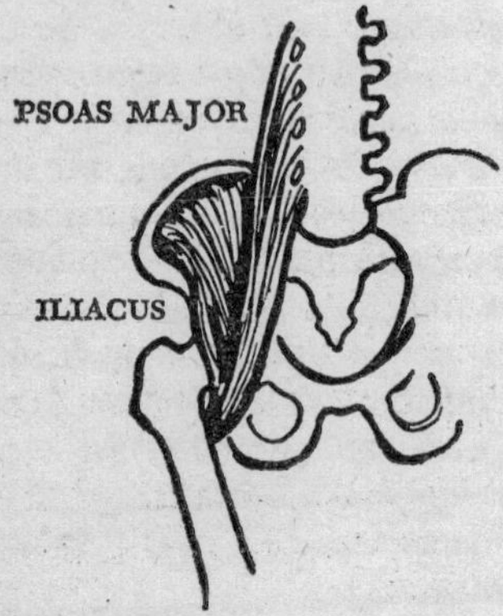

area. You can see, therefore, that it is one of the main controllers of posture in the body. Because the iliopsoas affects movement of the body in such a wide area, it is the primary source of most muscle imbalance, even though other muscles may also be functioning improperly.

Imagine what would happen if (for any one of a variety of reasons) this iliopsoas muscle was shorter than it should be. It could pull everything else out of kilter. It could cause a person to be round-shouldered, have a slumped posture, or sway-back. Or the short iliopsoas could pull on the femur – the thigh bone – where it attaches at the hip. When this happens, the abnormal strain on the thigh bone causes the upper leg to rotate outward. The lower leg turns also to compensate, twisting the two lower leg bones – the tibia and fibula – out of their proper relationship. The pulling produces a torsion effect, an outward twisting. You can see what stress this outward torque of the leg could create for the other muscles in the leg and foot. You may, at some time, have felt pain on the inside of the knee and outer hip and thigh from such torque or twisting. This torsion may also prevent the foot from resting squarely on the ground. The resulting rolling of the foot during walking creates another source of stress and pain.

The iliopsoas can also be too short on just one side. Then it will throw things out of kilter on one side and will cause many of the muscles on both sides of the body to change this way and that in compensation for the one-sided pull.

The iliopsoas is just one muscle, but it affects a whole network of other muscles. The iliopsoas and these other muscles are interdependent, like the rowers in a crew race. If the iliopsoas muscle is improperly developed or weak, all of the other muscles on the team must work too hard to compensate so that there is a constant possibility of strain at every point.

In fact, a short iliopsoas, by causing imbalances in all the other connected muscles, can result in every one of the following discomforts:

Flat feet, bow-legs, knock-knees, weak ankles, juvenile heel or instep complaints, cramping pain in the arch of the foot . . .
Tripping and stumbling in children . . .
Tilting of the pelvis to one side . . .
Distortion of the hip in newborn infants, often called congenital dislocation of the hip . . .
Hip pain and limping, especially in young boys . . .
Pain in the spine, legs, knees and feet in children, often called 'growing pains' . . .
Pain in the chest . . .
Weakening and subsequent fracture of the thigh bone of

older people, often mistakenly blamed on a fall . . .
Fractures or muscle ruptures occurring in army recruits, skiers, and basketball players . . .
Arthrosis of the knee . . .
Arthrosis of the hip . . .
Circulation problems . . .
Poor functioning of internal organs . . .
Fractures of the spine, or degenerative disorders of the spine . . .
Pain, tenderness or stiffness of the spine . . .
Herniated (slipped) intervertebral disc . . .
And plain poor posture . . .

It is almost unbelievable that a problem with this one muscle can be the hidden culprit that directly or indirectly leads to such varied ailments as dislocation of the hip in the newborn baby; a toddler's bow-legs, knock-knees, or difficulty in learning to walk; growing pains and awkwardness in older children; back or chest pain in young adults; arthrosis of the hip or knee in middle-aged men and women; fractures of the leg or hip in older people; and foot problems throughout life.

The case of a twenty-seven-year-old woman who came into my office recently is a perfect example. The wife of a medical resident at one of my hospitals, she had been plagued by a series of symptoms during the three years of their marriage. Finally her complaints became a sort of family joke.

For years she had been bothered by severe aching of the knees, hips and lower back. During her menstrual period the pains were often so severe as to be disabling. A gynaecological problem was suspected, but a thorough examination drew a blank. Her husband began to tease her about being a hypochondriac.

After a bout of spring cleaning, she developed a pain like a severe heartburn. Her husband sent her for a complete series of gastro-intestinal X-rays and gall-bladder tests. Once again she suffered the embarrassment of being told she was perfectly healthy. The doctor had no idea why she had such pain. Her husband began to call her 'the lady of the mysterious diseases'.

Then, during the third year of her marriage, she began to have headaches. The first few headaches were not too severe, and she thought they were probably sinus headaches, since she had some sinus trouble when she was younger. But the headaches became more frequent and much worse, until one day after a trip to Washington she was literally carried out of a party in acute distress. It was in Washington that her her husband suddenly remembered a lecture I had given at the hospital, which he had attended several months before. For the first time the 'mysterious diseases' began to fit together into a sensible pattern. His wife, who quickly became my patient, presented a classic case of a shortened iliopsoas muscle with secondary adaptations of malposture.

Her case history was a perfect progression of the symptoms which can result from muscle imbalance. First, the knee and hip pain came as a result of the uneven distribution of her body weight. The symptoms of heartburn were caused when the oesophagus, shortened from years of slumping posture, pulled the stomach up against it and partly through the diaphragm opening, making it possible for acid from the stomach to splash back up into the oesophagus. (When food is eaten, it goes through the oesophagus before entering the stomach. The diaphragm is the partition between oesophagus and stomach.)

When I took a detailed medical history of the young woman, I learned that she had always had trouble with sports when she was a child and thought of herself as clumsy. The headaches which finally brought her into my office usually came after she had carried heavy packages or had worn a heavy winter coat, putting an extra strain on the already strained muscles in her shoulders and neck.

This young woman is on a rigorous exercise programme and is making good progress. Most of her painful symptoms are gone. If her problem had been diagnosed when she was a child, the chances are that it could have been solved.

Prevention Is Possible. The shortening of the iliopsoas, which contributes to all of these problems, can be treated and the problems can be prevented or corrected if discovered early enough. It is disheartening to think of children and adults who could be helped if they could learn *not* to accept pain as a normal part of life.

A baby is born with his feet turned in and friends, or sometimes even doctors, say, 'Oh, that's nothing to worry about. It's part of normal development. By the time he's ready to walk, his legs will straighten out.' Nothing could be further from the truth!

Thirty per cent of all newborns have what doctors call the tibial torsion syndrome, some affecting one leg, some both legs. Which of them will correct themselves is difficult to predict. This syndrome consists of a combination of bowing of the lower leg bones, over-rotation or excessive turning-in (torque) of the lower leg and foot and a high instep. When the child begins weight-bearing by walking, the bowing deformity may correct itself. The rotational or torque deformity remains, and the child must adapt or adjust by developing abnormal compensation at his knees, legs, ankles or feet.

So the child begins to walk; his legs still turn in, he stumbles and falls and he can't understand what's wrong. If he turns his feet out, he gets flat feet. If he tries to roll on his feet as he walks he gets a condition called jack-knifing of the knees. Or he becomes bow-legged or knock-kneed.

But the parents are still being told by their friends and family 'he'll grow out of it'. And by the time he is five or six years old their child is seen by the paediatrician only once or twice a year, and never during an activity which might show up a postural irregularity.

But he does *not* grow out of it. He simply adjusts in a different way and the stumbling toddler grows into the eight- or nine-year-old child with back pains and hip pains, who can't sit up or stand straight and who is awkward, self-conscious and miserable.

When a child stumbles and falls frequently, or when a

child complains of pain, *it is not natural!* It is not part of growing up! It is nature's signal that something is wrong, and that something needs to be done.

If nothing is done, the child continues to grow and the problem continues to grow along with him. And soon the child with posture and walking problems grows into the adult with back pain, flat feet, bunions and arthroses.

In addition to these conditions *caused* by the poorly functioning muscle system, there are many disorders not initially caused by muscle imbalance, but made worse by it. For example, muscle imbalance obviously does not cause asthma; but if a person has asthma and suffers from poor posture and poor muscle function, these can make the asthma worse and can actually hinder treatment. The conditions that muscle imbalance can contribute to in various ways include asthma, emphysema, cerebral palsy, and varicose veins and other circulatory problems.

Just as the painful symptoms *caused* by muscle imbalance can be cured or at least significantly improved by proper treatment, many of these serious disorders aggravated by muscle imbalance also can be improved.

You Don't Have to Hurt. You, too, may have accepted your daily aches and pains and your physical hang-ups as a part of life you had to learn to live with. But they are keeping you from leading a full, happy and active life, and *you don't have to accept them.*

Your feet do not have to hurt after a long day of shopping.

Your back does not have to hurt after long hours in a car.

You do not have to come home from a tense day at the office with a headache.

You do not have to suffer pain in your neck and shoulders after typing twenty pages.

You do not have to limp home after a day of waiting on tables.

And most of all, you don't have to be known as 'the clod'. Your clumsiness at sports and dancing is not a permanent trait.

Neither pain nor awkwardness need be inevitable parts of life. They are a sign that the body is being mistreated, and something should be done about them.

What About You? Are you as physically fit as you could be, or is there a possibility that you might be among the thirty per cent of people who have some degree of musculo-skeletal imbalance? It it possible that you could be feeling better right now, that you are needlessly suffering from back or hip pain, sore feet or any of the other dozens of ills man is heir to?

What about the other members of your family? Could your wife or husband be enjoying more happiness at home or more success at work if she or he were feeling better? Would their personality be better, their days happier, their nights sexier if they weren't always besieged with nagging headaches, backaches, and other nerve-racking aches and pains that day by day tear away at the body and at the emotions?

Could your child be among this group with muscle problems? Could his rigid muscles and strained joints be causing his inability to concentrate on his studies or to withstand long car trips? Could he be suffering from self-consciousness, embarrassment and guilt because he is not good at sports or dancing, because he can't seem to keep up with the other kids or because kids call him clumsy or even a sissy?

It is simple to find out whether you or other members of your family are among those who have some weak or shortened muscles or musculo-skeletal imbalance. We have designed a self-testing quiz and a series of physical tests that will tell you in a few minutes whether you are among the one out of three with this problem and whether the exercises in this book can help you.

First do the quiz. Then try the physical tests in Chapter III. And, if you are among those who suffer the results of muscle imbalance, plan to start the curative exercises that begin in Chapter IV, immediately.

SELF-TESTING QUIZ

Read the following questions and score yourself in the blank spaces provided in the right-hand column. Follow directions for finding your Personal Score. Then study the Scoring Guide to determine whether you have muscle imbalance and what to do about it.

SYMPTOMS

1. Back pains. *Enter the average number of times per month (times two) that you experience significant backache.* ______

2. Sore feet. *Enter the approximate number of times per month (times two) that you experience aching feet (excluding normal adjustment to the stiffness of new shoes).* ______

3. Hips. *Enter the number of times per month (times two) that you are aware of pain in the hip region.* ______

4. Neck or shoulder blade tension. *Enter the number of times per month (times two) that you get a 'crick' in your neck or pain between your shoulder blades.* ______

5. Headaches. *Enter the number of times per month. Multiply by two for severe headaches.* ______

6. Cramps of leg muscles. *Enter the number of times per month (times two) that you get cramps or muscle spasms of feet, lower or upper legs.* ______

7. 'Pins-and-needles'. *Enter the number of times per month that you experience this type of muscle cramp in thumb or fingers.* ______

8. Spinal flexibility. Do you have trouble straightening up again after being in a bent-over position? *Enter 5 for often, 3 for occasionally.* ______

9. Chest pain or heartburn. *Enter the number of times per month (times two).* ______

MEDICAL HISTORY

Enter 3 points in the blank at the right for each of the following conditions you now have or have had at any time in the past.

1. 'Sway-back' or curvature of the spine. ______

2. Round shoulders or 'rainbow' curvature of the upper spine. ______

3. Dislocation of the knee cap. *Enter 3 points for each dislocation.* ______

4. Stress fractures of the hip, thigh, leg, ankle or foot. *Enter 3 points for each.* ______

5. Pulled or ruptured muscle, including 'charley-horse' strain of the calf muscle. ______

6. Bursitis of the shoulder or hip. ______

7. Sciatica. ______

8. Arthritis or rheumatism. ______

CHILDHOOD EXPERIENCE

Enter 2 points in the blank at the right for every one of these conditions you experienced at all as a child.

1. Did you have trouble learning to walk? ______

2. Did your feet turn in? ______

3. As a toddler did you walk on your toes for a long period? ______

4. Did you stumble or fall over your own feet? ______

5. Did you have 'growing pains'? ______

6. Did your parents often tell you to 'straighten up'? ______

7. Did you have trouble sitting still in school, or long enough to get your homework done? ______

8. Did you avoid sports participation as a child because of poor coordination? or because the team captain never chose you for his team? ______

YOUR PERSONAL ANATOMY

Score 2 points for each of these conditions that apply to you.

1. Do you have trouble crossing your legs to put on shoes? (Do not count if difficulty is due to obesity.) ______

2. Do one or both feet turn in or out when you walk? ______

3. Do you have bow-legs? On one or both sides? ______

4. Do you have knock-knees? On one side or both? ______

5. Do you have flat feet? ______

6. Do you have bunions? ______

7. Do you have 'pins and needles' in the front of your feet or your toes? ______

8. Do you have painful heels or spurs? ______

9. Do you have calluses, hammer toes or loss of sensation in the middle toes? ______

10. Do your knees bend backwards (jack-knife)? ______

11. Is one of your legs somewhat shorter than the other? ______

12. Is one shoulder higher than the other? ______

13. Are you round-shouldered? ______

OTHER CLUES

1. Are you restless when you have to sit in one place for an hour or longer? *Enter 2 points for frequently, 1 for occasionally.*

2. Are you excessively uncomfortable or do you get neck, shoulder or lower backaches during long car rides? *Enter 2 points for frequently, 1 for occasionally.* ______

3. Do you slouch in a chair when you sit? *Enter 2 points.* ______

4. Do you have a 'peculiar' gait? Do others comment on the way you walk? *Enter 3 points.* ______

5. Are you awkward or inept at sports or dancing? *Enter 2 points.* ______

6. Do you trip over your own feet, or stumble? *Enter 2 points if frequently, 1 if occasionally.* ______

7. Have you sprained your ankles often (*enter 3 points*) or occasionally (*enter 1 point*)? ______

8. Do your ankles get tired and 'wobbly' after long stretches of walking, skating or skiing? *Enter 2 points.* ______

9. Do you get bunions, corns, calluses or 'pump bumps' from your feet rubbing against your shoes? *Enter 2 points for frequently, 1 for occasionally.* ______

10. Do your heels wear down on one side of your shoe rather than evenly? Or does one shoe wear out faster than the other? *Enter 1 point for each irregular shoe.* ______

11. Do you buy wedges or pads for your shoes? *Enter 2 points.* ______

12. Do you have difficulty getting shoes that fit? When the heel fits, the front of the shoe is too small or vice versa? *Enter 3 points.* ______

TOTAL .... ______

YOUR PERSONAL SCORE

1. If you are twenty years of age or under, *add 15 points to the total above* and enter result here. ______

2. If you are between 21 and 35 years, *add 10 points to the total above* and enter result here. ______

3. If you are over 35, the score totalled above is your personal score. Enter it here. ______

SCORING GUIDE

If your score is 20 points or less, you are remarkably free of the symptoms of muscle imbalance, and are probably enjoying good overall health and are physically fit.

If you scored between 21 and 30 points, you may be among the lucky two-thirds of the population who do not suffer from muscle imbalance, but you should try the physical tests in Chapter III to see whether you have a hidden muscle imbalance.

If you scored between 31 and 45 points you are very likely to have some degree of muscle imbalance and should definitely take the tests in Chapter III.

If you scored between 46 and 60 points, you have almost certainly lost many working hours or even days due to the painful symptoms of muscle imbalance. You should probably get professional advice about adjusted footwear, and should embark immediately on a programme of corrective exercises.

And if your score was 61 points or more, you probably don't need me to tell you that you have spent many more days than you like to remember in extreme discomfort and have already consulted a number of physicians with little or no relief from your painful symptoms. You should definitely seek professional advice about adjusted footwear.

You absolutely must start doing corrective exercise without delay, and should resolve to stick with the complete programme that is outlined in this book until you experience the truly blessed freedom from pain that your body is capable of attaining.

CHAPTER TWO

How Your Bones and Muscles Work

To understand muscle imbalance, and to understand how exercise and other treatments can help, you first need to understand the normal structure and motion systems (kinesiomechanics) of the body. If you know how your muscles and bones work normally, then you will better understand what has happened when they are not working normally.

The Body's Engineering System. What is the physiology of motion? What are the forces and actions that allow us to walk and reach out and turn and hold on to things? How do our muscles and bones act together to establish such perfectly balanced forces that we are stabilized and in equilibrium, that we don't fall on our noses or trip over our feet?

The anatomy of your musculo-skeletal system is relatively simple to understand. Think of the bones as being the framework on which your body hangs together. Yank them out and you would be a human jellyfish. The more than 500 muscles in your body are draped around the framework of bones.

The muscles contract and relax and act as springs and pulleys, moving bones up and down, back and forth. The bones are the levers, and the joints are the fulcrums or pivot points about which the bones move. How and where a muscle is attached to a bone, the shape of the muscle, the shape of the bone it is attached to, and the kind of joint involved – all of these determine what kind of action will result from the contracting and relaxing of a certain muscle.

When the muscles are in normal balance, they make the

body work smoothly; when they are out of balance, there is stress and strain on the joints. It's all a matter of the basic principles of physics.

Kinds of Joints. There are several types of joints. Bones can be joined together in a way that allows them to move, or in a way that stabilizes them.

A *hinge joint*, like the hinge on a door or a trunk lid, allows the muscle to pull the bone in only one direction. Your knee joint is a good example. You can bend your knee or straighten it, but you can't bend it backward (hyper-extend), rotate it or move it in any other direction.

A *saddle joint* is shaped somewhat like a western saddle with two ends tipped up making hollows in the surfaces in both directions. This allows forward and backward as well as sideways movement. An example is your thumb joint, where the bone at the base of the thumb is attached to the bone of the hand. Another is the hindfoot – the joint between the talus and the calcaneus. Abnormal deflection of this joint inward results in what is called a varus heel, while an outward deflection results in a valgus heel (flat foot).

A *ball-and-socket joint* allows the round head of one bone to fit into the socket of the other bone, like the swivel joint on the top of a camera tripod. It permits bending, straightening, forward or backward motion, or rotation – just about any kind of movement you may want to make. Examples are the shoulder and hip joints.

Still another kind of joint connects the vertebra in the spinal column. The vertebrae are joined together by cartilage that touches and connects each vertebra to the next. The result is a supple chain that allows the motion of twisting and bending.

What Is a Muscle? What Does it Do? A muscle is made up of tiny fibres or cells, each stretching the length of a bundle of muscle tissue like a thin thread. These fibres are joined to make larger bundles, until a complete mass of muscle is formed. Each of the long, skinny cells is capable of contracting when stimulated by a nerve impulse. And when a group of muscle fibres contracts, the muscle mass they make up contracts and shortens.

The attachment of a muscle determines what kind of work that particular muscle will have to perform. Muscles and their attachments determine the positions and movements of all the various parts of the body in relation to each other.

If you looked at a piece of muscle under a microscope, you would see that each bit of muscle is made up of a bundle of fibres, each about the thickness of a human hair. But each hair-thin fibre can support as much as a thousand times its own weight. There are about 6 *billion* of these tiny muscle fibres in your body.

If you were to examine many muscle fibres under the microscope, you would notice that there are three kinds. One has dark and light stripelike bands across the sheathed fibres and is variously called striped muscle or skeletal muscle (because it is attached to bones of the skeleton) or voluntary muscle (because it is used for voluntary movements such as throwing a ball or walking).

A second kind of muscle fibre is smooth and not striped. It is called involuntary muscle because it handles all the automatic involuntary work of all the internal organs (except the heart). It controls such things as breathing, stomach action and intestinal movements.

The heart has the third type of muscle, called cardiac muscle. These fibres are striped, but are not separated from each other by sheaths as the skeletal muscles are. Instead they are joined in a continuous network. The heart muscle is also involuntary muscle, operating without a signal from the brain.

There are two entirely different nervous systems for the voluntary and involuntary muscle systems. In the case of the voluntary muscles, the signal is sent from the brain through the *central nervous system.* The split-second signal races along nerves from the brain to the spinal cord, where the signal is relayed to another nerve that shoots it to the proper muscle fibres. Thus stimulated into action, the fibres

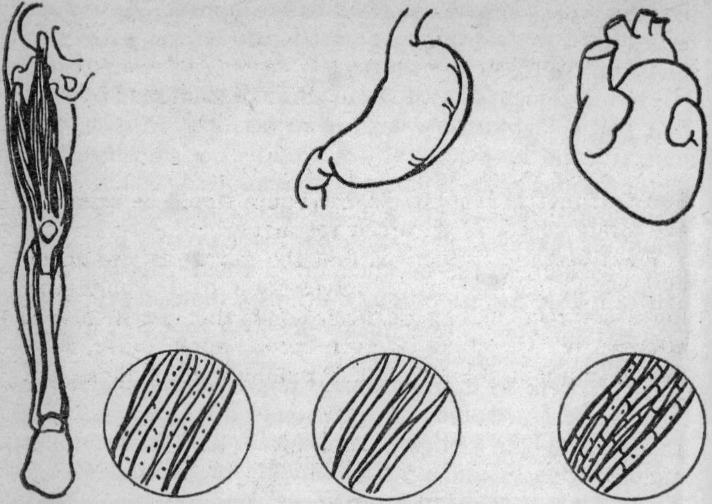

VOLUNTARY MUSCLE INVOLUNTARY MUSCLE CARDIAC MUSCLE

all contract simultaneously, becoming almost a third shorter and correspondingly thicker.

The smooth, involuntary muscles, however, receive their signals from a different set of nerves – the *autonomic nervous system*. They are 'automatic' and do not need to be 'thought' about and directed. These nerves originate in a different part of the brain. They are divided into two sub-systems: the sympathetic nervous system and the parasympathetic system. These balance and counteract each other. The sympathetic dilates the pupils of the eyes; the parasympathetic constricts them. The sympathetic nerves speed up the heart; the parasympathetic slow it down.

Muscle Attachments. Muscles also differ in how they are attached. In the following diagram of two bones and two muscles, for example, if muscle A contracts, the bones would move up as shown by the arrow, causing the joint to be

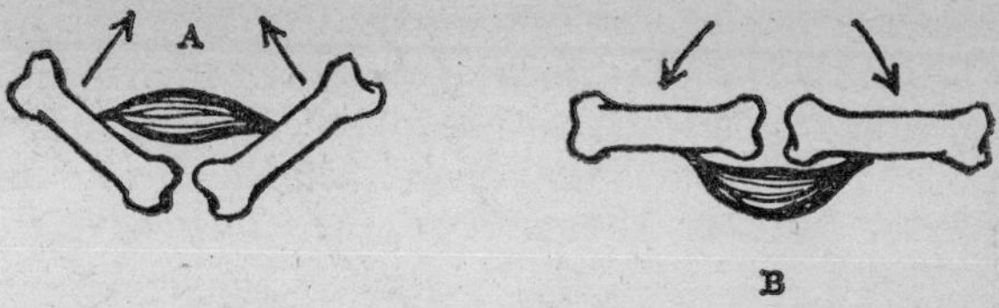

flexed. If muscle B contracts, the bones would be extended and straightened, as shown by the arrow.

What actually happens is that the muscle is stimulated by a signal from the spinal nerves going to it. The muscle fibres contract, pulling on both bones that the muscle is attached to. Theoretically, both bones should move, since the tension is in the middle and pulling both of them. But

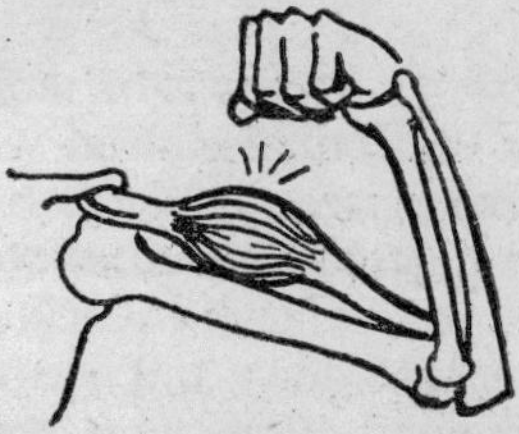

usually one bone or the other is stabilized, held motionless by other muscles, so just one of the bones will move when the muscle contracts. If you kick your leg out, the upper part of your leg remains still. Only the lower leg moves out when the big muscle contracts.

The point where the muscle attaches to another bone that holds steady is called the *origin*. The point where the muscle attaches to the bone that moves is called the *insertion*.

So what a muscle does – whether it flexes a leg or extends it, whether it moves a bone up or down, backward or forward, or causes it to rotate – all depends on where the muscle is attached, how it pulls, and how the joint lets the bone move.

This explanation is oversimplified, of course. Most muscles have many attachments, and many muscles work together or against each other. And some muscles simply act as stabilizers, steadying and supporting various parts of the body. Some act as brakes, to keep the most powerful muscles from moving a bone too far. And still other muscles are more or less 'emergency muscles', giving help to other muscles when extra force is necessary or when they have to act against resistance.

Usually muscles are at least paired in their action. That is, for one muscle that causes your arm to flex, there is at least one muscle that does just the opposite, making your arm extend. So when one muscle contracts, another one must relax. In fact, your nervous system is set up with an elaborate switching system so that when a nerve impulse goes to one muscle to make it contract, an impulse also goes to its opposite action muscle to inhibit its contraction and cause it to relax.

Some muscles extend over two or more joints, so other sets of muscles also become involved when they act. For example, some of the muscles going along your fingers also go along your wrist. Hold your arm, wrist and hand out straight. Then flex your fingers to make a fist. Open the fist. Now bend your wrist down and try to make a fist. What do you feel happening? When you bent your wrist the muscles that cover the several joints had to stretch out, so the muscles could not contract to form a fist as before.

The same principle is operating in this example: stand up and raise your upper leg as far as it will go. Let your knee bend. Now hold your leg in the same position and try to straighten your knee out. You can't do it without lowering your leg, unflexing the hip joint.

One Muscle Can Affect Your Entire Body. By now you are aware of how muscles interact. When you realize that muscles may be attached to as many as a dozen different bones as well as to other muscles, you can begin to understand how one muscle can affect many parts of your body. And you can realize how one muscle's being too short, too rigid, too weak, or otherwise improperly balanced in relation to other muscles can upset the functioning of the other muscles. As one muscle functions abnormally, other muscles all down the line must compensate and make adjustments.

Sometimes you can actually feel your muscles pulling in their many ways, interconnecting in their complex systems throughout your body. If you stretch your arms high while standing on your toes for a long time, as when putting things on a high shelf or wallpapering a ceiling, you can feel those muscles pull right from your fingertips down through your shoulders, your back, and down into your hips and legs, and even into your straining arches and toes.

And it's no wonder you can feel the stretch all over. In medical drawings or diagrams of the hundreds of muscles of your body, you can see the astounding number of interconnections and interactions. When you carry a heavy shopping bag of groceries for ten minutes, not only will your arm and shoulder muscles get tired, but your neck also tightens and your back hurts. In the same way, many muscles are affected when you bend over while gardening or when you work at the sink for a long time. It even takes a lot of muscles working together throughout your body just to keep you balanced when you stand up.

One test you can make at this very moment, to prove to yourself how these muscles are all tied together, is to tighten and then relax your buttocks. Even though you are aware of tightening only the muscles in one area, you will feel your arches pull up and your stomach pull in.

The iliopsoas muscle is especially important because it interacts with and influences so many other muscles and parts of the body. The iliopsoas is mainly a broad, flat muscle in the lower back, but like an octopus, it has arms reaching out in many directions. It has segments that go to every vertebra in the low thoracic (chest) and lumbar (back) areas of the spinal column, and other segments that go to the pelvic and thigh bones. It pulls in many directions, and it is powerful. In its normal function it makes the thigh move forward as well as rotate outward in the hip socket. It also causes the natural curve in the spine and controls the pelvic tilt and general posture.

Abnormal functioning of this one major muscle can have ramifications reaching through a kinetic chain reaction up to the back of your neck and down to your big toe. And of course there are other important muscle systems as well, in addition to the iliopsoas, all with their own interrelationships and interconnections, and with their own symptoms when something goes wrong.

Because there are so many interconnections, it sometimes takes a while to find out which muscle is causing a person's problems. Because so many different muscles may be compensating, in order to determine which one is primarily at fault, it may be necessary to relax the muscles in that system one by one.

Thus it may be discovered that a painful back is not really a bad back at all, but a referred symptom of improper functioning of the spino-pelvo-femoral muscles of the hip. And pain in the hip is frequently determined to be due to the fact that the iliopsoas muscle going to it is so rigid that the thigh bones can't rotate properly without grinding in the hip socket.

This is precisely what happened in the case of Judy R., a pretty woman of about twenty-five who had recently been married. Her problem, she said, was a severe pain in her shoulder radiating into her hand, a one-sided headache, and a miserable backache that simply would not go away, and was so bad that sometimes she couldn't work. The pain became worse when she had to sit for very long, when she stretched to reach for something, or when she drove a car. Sometimes the pain was sharp and severe, sometimes it was dull, but nagging and constant, a depressing and grim overtone to everything she did. Because of it, she was nervous and tense and often snapped at her husband.

She had consulted several doctors, but none of them had been able to pinpoint the cause of her backache. Pills and other treatment had provided only temporary relief. At first she had been told that she was tense and nervous because she was twenty-four and still single. Then she was told that she was tense and nervous because of the strains of being married.

When I put her through an extensive series of muscle-function tests, I determined that muscle tension was causing severe strain on her joints and not the tensions of her motions or the strains of her being single or married. Ultimately I was able to trace her various symptoms, head and back pain, to a rigid muscle in the hip.

I guided her through a programme of exercises, and she did them diligently and with enthusiasm twice a day for half an hour at a time. She continued to come to the office every other week for therapy, and at each visit both she and I could see how much progress she had made toward achieving muscle flexibility. After six weeks of the exercise programme she was able to report that for the first time in two years she wasn't in pain.

Treatment to Restore Normal Engineering. Knowing what forces are involved, and understanding how muscle disturbances can occur, one can see the logic of using exercises as well as other techniques to restore the muscle balance to normal. Muscles that are short or rigid can be corrected by exercises specifically designed to help them – the orthotherapy exercises. Orthotherapy can stretch these misaligned muscles, relieving the abnormal strains that have been pulling on other muscles and bones.

And, because other related muscles have over a period of time developed abnormal function because of adjustments and compensation, a wide range of painful symptoms may be relieved by exercises designed to return these muscles to their proper tone and function as well.

Therefore a corrective exercise programme must first be directed toward a restoration of normal function of those misaligned muscles that have adapted or adjusted to the primary malfunctioning muscle. Later, after there is some relief of disturbing symptoms and the adjusted muscles have been relaxed, exercises are added for correction of the primary culprit, which in most cases is the iliopsoas.

CHAPTER THREE

Physical Tests for Adults

If you took the quiz at the end of Chapter I, you already have some idea of whether you have muscle imbalance.

Now in this chapter we will give you the physical tests to determine whether or not your daily aches and pains are really due to this problem – and whether you are among the many whose muscle imbalances may be hidden at the moment but are likely to cause pain and disability later in life. (Physical tests that you can perform to judge whether your child has a muscle problem can be found in Chapter VI.)

The many signs and symptoms of muscle imbalance can occur at all ages and in many forms. Musculo-skeletal imbalance is one of the biggest masqueraders and mimickers in the entire field of medicine. A doctor must be a bit of a Sherlock Holmes to detect it, interpreting clues that often seem to be something else. To make it even more confusing, the signs of muscle imbalance are different at different ages; and these disorders work on your body progressively over the years. Thus, symptoms spread from one part of the body to the other through the kinetic chain, as stress and strain are projected on to more and more muscles and joints.

In the last chapter we saw how the muscles are interconnected and learned that, in a chain reaction, what affects one goes right on down the line and affects the others.

A frequently occurring example of this chain reaction begins with the hip and back muscles that are tight and out of line. They pull on the thigh bone and cause it to rotate.

This in turn puts the knee under a strain, which causes poor posture, which affects the way you walk, and affects the muscles, bones, joints and ligaments of spine, pelvis, hips, knees, ankles and feet.

So one must be alert for signs and symptoms all along the line. Your muscles or those of your child may even now be at one of the stages in this chain. You want to halt the damage right where it is now, before it spreads further. And you want to alter the series of symptoms – which can be done.

In this chapter you will learn how to detect the physical signs of musculo-skeletal imbalance.

Physical Signs to Look for: Infants and Children. See Chapter VI for specific physical tests that you can make to detect hidden muscle imbalance in your infant or child. Even in the *newborn*, you and your paediatrician should be on the alert for indications of such abnormalities as wry-neck (torticollis); congenital abnormal curvatures of the spine and rib cage; depressed (funnel chest) or elevated (pigeon breast) sternum or so-called breast bone. A lump in the lower spine denotes a spina bifida vera, which demands immediate medical attention.

If the infant's hip joint remains flexed, or bent, so that his legs do not straighten out, congenital hip dislocation may be indicated. Only your physician can make this, or any other, diagnosis. But it is up to the parent to 'take inventory', as it were, of the infant's appearance and call any possible problem to the doctor's attention.

Parents should be particularly alert for changes that occur when the child begins to walk; because of malalignment of the bones of the leg and foot he may be placing weight and therefore greater strain on certain muscles. When the child's heel hits the floor in walking, you may notice that his foot rolls inward. This is the most frequent cause of stumbling in children, and is often the cause of many falls for the toddler. Many toddlers insist on walking on their toes most of the time. This often demonstrates the need for a heel pad to adjust for the high-arched instep of the infant with tibial torsion syndrome (see page 10).

Sometimes, instead of the inward rolling of the child's foot, or perhaps in addition to it, you will notice that his pelvis rotates out to the side or backward when he walks, making the buttocks seem abnormally prominent. You sometimes see this same kind of rotation in a young girl with a big 'rear' walking on high heels she's not quite used to yet.

Watch to see if the child turns his toes in or out. If your child's feet are not parallel, check to see if his legs are turned at the hip, or only below the knee as in tibial torsion syndrome.

Look to see if the child is knock-kneed, bow-legged or holds his knees bent backward (jack-knifed). Is one shoulder higher than the other? None of these is normal.

'Growing pains' are the accepted complaints of a child during his rapid-growth years. In reality these complaints are Nature's earliest signals of postural distress. Unless you are aware of the normal and abnormal kinetic chain reactions of the erect posture of the spine and extremities and can relate them to musculo-skeletal imbalance, 'growing pains' will be impossible to interpret.

Be alert for complaints of pain in the hips, knees, legs or feet. As a child who has this disorder gets a little older, he is particularly apt to get 'recoil' pain when he gets into bed at night. He has stretched his muscles abnormally in active play all day, and finally when he's off his feet for an hour or two, he develops cramping pains in the arch of the foot or the calf of the leg. This, too, is part of the same syndrome.

At the age of seven or eight, a child with a muscle problem may still be falling a lot. This is the age, too, at which you may start noticing that he's not as athletic as some of his companions. He can't throw a ball well, run normally, skip or jump rope, play hopscotch or play with reasonable skill other games his playmates seem to be able to handle with ease. He may hang back and say he doesn't want to play, because he himself is aware of his lack of ability, of his clumsiness in comparison with the other kids.

At about this age, too, you no longer bathe your child, but let him do it himself. He becomes selfconscious about his body and dresses alone in his room with the door closed, or hidden in the bathroom so you don't see him naked any more. Unfortunately it is also at about this age that many physical signs of muscle imbalance start showing up, and so they often go undetected. A slight humping of the dorsal spine or rounding of the shoulders; a little 'pot belly', especially in males; in females, abnormal curvatures of the dorso-lumbar (lower) spine resulting in *scoliosis* – it is easy to miss these indications unless you are watching for them. Recognition of spinal disorders at this early stage of development presents the golden opportunity for correction by prescribed exercises.

You may find yourself constantly telling your child to stand up straight. You may notice that he's always twisting and turning, especially when seated, in an attempt to relieve tensions and strain on his spine. He may want to start riding to school instead of walking. He may not be able to sit still in class, fidget while watching television, squirm around while doing his homework or reading.

In years gone by, it was often the dressmaker who detected irregularities of the spine and shoulder during a fitting. Today, it is entirely up to the family to be alert to the earliest signals of spinal and shoulder irregularities and seek correction through early management.

By the time the child is nine or ten, a noticeable limp may appear. He may complain of pain on the inside of the knee; you now know that this indicates that the muscles are putting unusual strain on the inner side of the knee joint. This is likely due to disequilibrium of the hip, and dynamic testing for disorder of the hip joint should be undertaken by your physician.

All of these symptoms in children aren't just 'growing pains'. They mean something, and you should see your doctor to find out what to do. If he says 'Don't worry, they're just growing pains,' or 'Don't worry, he'll grow out of that funny walk,' or 'Don't worry, the way his foot turns in will change when he grows up,' you know he's not right. But don't lie awake at night worrying – just get another medical opinion.

Physical Signs to Look for: Teenagers. Because the symptoms of muscle imbalance develop progressively, parents of teenagers should be alert for the same indications we have listed for young children. As it becomes more important for the adolescent to participate in such activities as sports and dancing to keep up with his friends, you may find other signs such as a tendency to turned ankles or dislocated kneecaps (patellae); or you may simply hear complaints about weak or sore ankles or pain that prevents your youngster from walking, running, skating or skiing well.

Physical Signs to Look for: Adults. In adults, the signs can vary a great deal, depending on how fast the secondary adjustments or adaptation effects have spread to other muscles and joints. An adult with muscle imbalance will probably recall that, when younger, he experienced many of the symptoms described above.

Even in young adults, there may be pain in the hips, or in the neck, shoulder and hands. By now the feet may be showing signs of fatigue and pain. There may be flat feet, hammer toes, corns or calluses. The weekend athlete may have trouble with strained or ruptured ligaments and muscles, and even occasional fractures. Arthrosis may have started to settle in previously injured joints, especially at the hips and knees. Sciatica, headaches, some apparent forms of indigestion and constipation, excessive fatigue – these are a few of the symptoms that show up in adults with this problem.

PHYSICAL TESTS

All of the previous symptoms can be readily noted, either by the person experiencing them or by a parent or other observer who is aware of their significance and is watching intelligently for indications of muscle problems. But because of variations in body chemistry and life styles, not all of us develop the same symptoms at the same age. Many people with muscle imbalance are fortunate enough not to be severely disadvantaged by painful symptoms. Instead, they experience occasional aches and pains, but for some reason have not yet developed the more disabling symptoms of this syndrome.

I recall one woman, for example, who for years put up with painful feet and attempted to mask her poor posture under well-chosen garments, but who was finally driven to consult me in her mid-thirties when she experienced in rapid succession the extreme pain of sciatica and what seemed to her regular physician to be a gall-bladder attack. Her medical history showed that she had experienced a long string of minor manifestations of her problem in childhood and early adulthood, but no one had noticed the interrelationship between her flat feet, bow-legs, weak ankles, problem posture and clumsiness at sports.

There are many like her. Muscle imbalance can remain hidden for quite a long time. Paradoxically, this hidden period is the best time to correct it, if you know how to detect it. When the effects of the weakened or deranged musculature are not yet so extensive or so severe, they are less resistant to treatment. Although muscle imbalance can be treated and corrected at any stage, the sooner it is detected and the less it has spread throughout the body, the easier it is to treat and the more effective the treatment can be.

For this reason, it became very important to me to develop physical tests that would unfailingly demonstrate even a latent muscle imbalance. The rest of this chapter, therefore, will describe tests that you can do yourself, or with help from another person, to determine accurately whether you have musculo-skeletal imbalance, even if you may have noticed only a few symptoms to date.

Because muscle imbalance is a progressive ailment, its symptoms vary according to the age of the patient. For example, the same stress will produce different hip conditions at various age levels, since the hip, like the other joints, changes with skeletal growth.

In the newborn, dynamic testing will disclose conditions called *congenital coxa vera* and *congenital hip dislocation*; in the infant, *Leggs-Perthe's disorder*; in the teenager, *slipped femoral capital epiphysis*; in the adult, *degenerative arthrosis of the hip*; in the aged, *coxae malum senilis*. The nature of the abnormal forces producing each of these disorders is one and the same. Therefore, dynamic testing for musculo-skeletal imbalance is the same for people of all ages, and the age of the person tested determines which disorder will manifest itself at which stage of hip development and growth.

These tests are of three kinds: General Posture; Muscle Tension; and General Physical Fitness.

1: POSTURE TEST

First, look at yourself in profile in the mirror. No cheating – stand the way you usually do! Now compare your profile with those of the man and woman in the drawing on

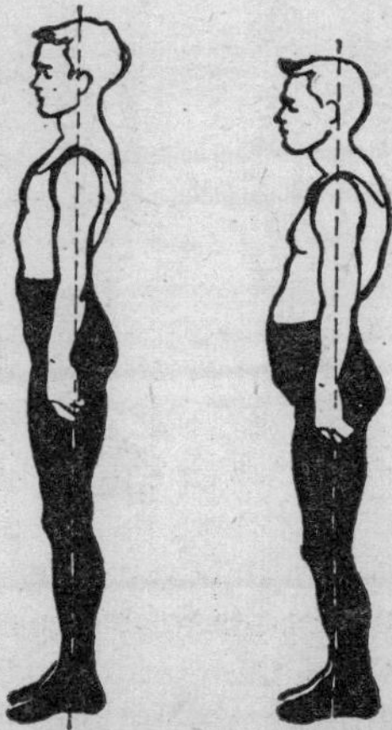

this page. Does it look like the good profile of the man, or the bad one of the woman?

Good posture is a result of holding your body in a correctly balanced position. The easiest way to attain it is to think of an imaginary line running (side view) from the top of your skull down through your neck, shoulders, hips and insteps. If the head is bent forward, the abdomen thrust out or the back stooped, the body is out of normal balance, and a strain must be placed on various muscles to keep the body from falling.

Here is another test: stand with your back against a wall, with your head, heels, shoulders and the calves of your legs touching the wall, your hands by your sides. Flatten the hollow of your back by pressing your buttocks down against the wall. You should just barely be able to stick your hand in the space between the wall and the small of your back.

If this space is any greater than the thickness of your hand, you have a posture problem.

In the same position, does your profile resemble that of

the person illustrated here? The 'ruler' on the left in the drawing corresponds to the wall behind you. If your profile is approximately the same, your posture is excellent and you can grade yourself 'A' on this one.

But if your heels do not touch the wall when your shoulders do, you should be graded 'B' or 'C'. If possible, have someone place a yardstick upright next to you, starting at the middle of the side of your foot. It should coincide with the dotted line of the centre of gravity on the figure in the drawing, touching the front of your leg below the knee and going straight up through the centre of your torso, neck and head.

Now stand facing the wall, right up against it, with the palms of your hands resting against the front of your thighs. If your chest touches the wall first, your posture is probably excellent (A) or at least good (B); if your head touches first, it is only fair (C); if your abdomen touches first, you get a 'Z' for zecch!

In testing a child's posture you must take into account the differences between the normal postures of adults and children, as well as the wider range of individual variations in body forms in children.

The pre-school-age child has a prominent abdomen, a little beer-belly that would be completely unacceptable at age forty. At ages six to twelve, there may be natural increased lumbar lordosis ('swayback') and slightly rounded shoulders. The chest is flat. In the adolescent years, girls develop feminine proportions and boys show muscle development. By mid-adolescence, the abdomen should be flat, and normal posture resembles that of an adult.

2: TESTS FOR MUSCLE TENSION AND MUSCLE IMBALANCE

Two people should work together to test themselves for signs of muscle weakness and imbalance.

The person being tested should either be nude or dressed in a leotard or similar form-revealing outfit that permits free limb movement. First, if you are the partner-observer, have the person being tested walk across the room from you, turn round and walk back. Is there any peculiarity about the way he walks? Is he pigeon-toed? Do his legs or toes thrust outward with each step? Does he limp? Do his hips swing back and forth with each step?

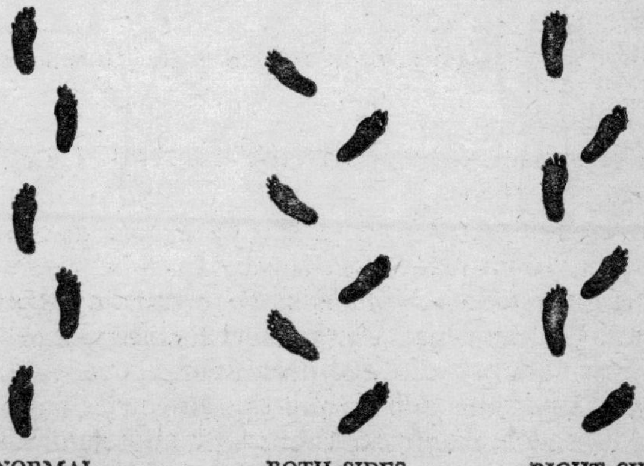

NORMAL — BOTH SIDES ABNORMAL — RIGHT SIDE ABNORMAL

Another way to check someone's way of walking is to analyse his footprints. Walking in sand gives beautiful footprints for analysis; or, you can get clear footprints by immersing the feet in water and then walking on newspapers laid out to form a path. Compare the wet or sand footprints with those illustrated, which show the difference between normal and two abnormal sets of footprints.

Now have the person stand erect and face you. Are his shoulders even, or is one shoulder higher than the other? Is he knock-kneed or bow-legged? Now have him turn around, with his back towards you. Check his shoulders again from the back to see if they are uneven. Is his spine straight, or is there a curve to one side or the other? Look at his seat. Are the two sides symmetrical, or are the muscles or creases of skin different on one side than on the other? Is one hip higher than the other? The illustration shows how this appears.

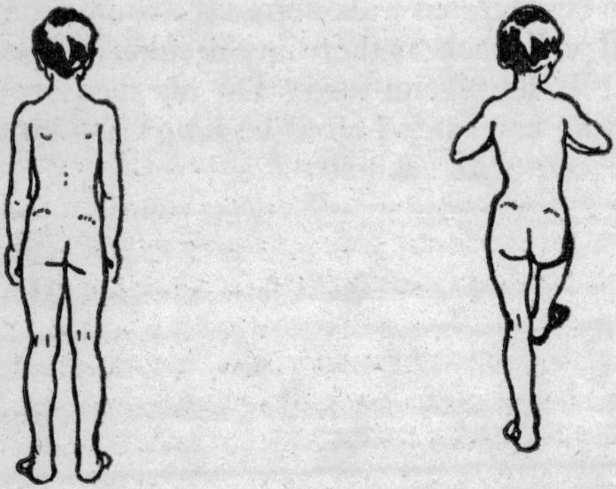

While he is standing with his back to you, have him raise one foot. Do both hips stay parallel to the floor when he does it, or does one tilt? See illustration above right.

Have the person stand up straight. Now tell him to bend over and touch his toes. Those with muscle imbalance usually cannot – in fact, they may not get their fingers much lower than their knees, if the problem is even moderately severe.

Now tell your subject to sit down and cross one leg over the other as though trying to put on a shoe. If he has pain or cannot do it, it is a strong indication that he is among those who need help.

Next, have him sit upright on a table or bench with his legs extended on the table in front of him. If he can't keep

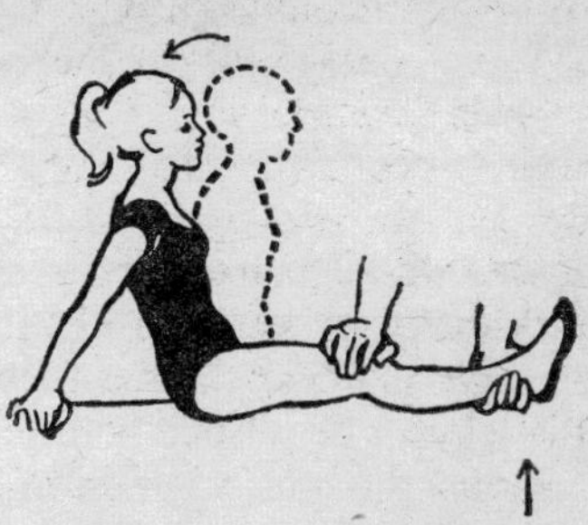

his legs fully extended while sitting up straight, or if it hurts to do so, muscle tension is indicated.

A variation of this test is for the subject to lie on his back on the floor with his legs extended and his feet pushed up flat against a baseboard. Now have him sit up slowly. If his legs bend at the knee joint because he can't keep them flat down on the floor, contracted or shortened muscles are indicated.

Next, have the person lie on his back on a table or bench and see if the small of his back has an excessive curve, called lordosis (see illustration).

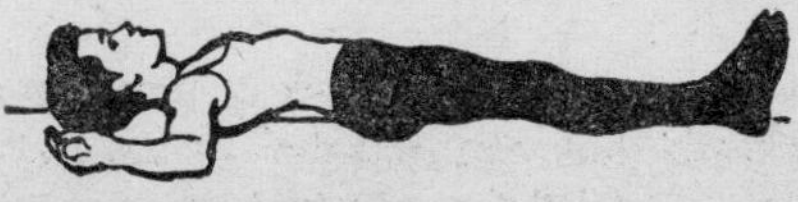

Now have him lie on his back on a table with his legs hanging over the edge. With one hand, press one of his legs, bent at the knee, up against his abdomen (see illustration).

Then hold the other leg just above the knee and push slowly down towards the table. Normally it should be possible to push this leg all the way down to the position of the dotted line or even lower. However, if the iliopsoas muscle is tight

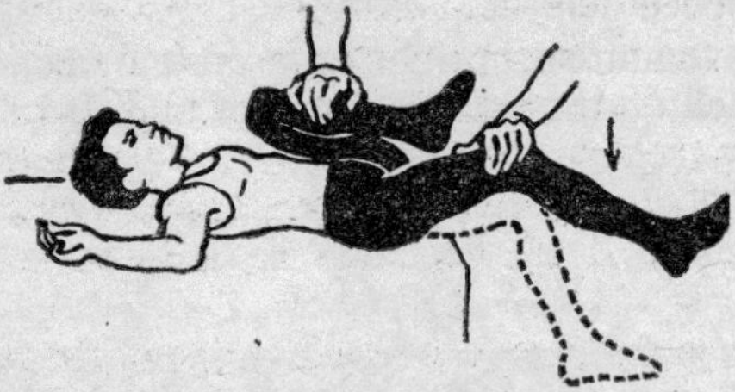

and tense, it will be very painful to push the leg down, or even impossible to get it lower than the level of the table. How far down the leg can be pushed before it starts hurting indicates how serious the muscle tension is. If the pain occurs when the leg is nearly down to the normal dotted line position, then there is only a little tension and contraction. But if the pain occurs when the leg is still considerably up and off the table, then the involvement is severe. Abnormality is also present if the person's other leg, being held against the side of the abdomen, bounces up when you push down on the first leg.

Here is another test: with the person on his back in the

same position, raise both legs together so that the knees and hips are bent or flexed, with the feet together, as illustrated. Testing the legs one at a time, stand at your subject's side and keep one hand on the knees to hold them firmly together. Then grasp the foot nearest you with your free hand, pulling the leg outward and towards you as far as possible. Then test the other leg. Normally it is possible to pull the leg outward to almost sixty-five degrees. If there is moderate muscle derangement, rotation may be possible to only about forty-five degrees, while in severe cases of muscle contracture, rotation of the leg may be reduced to five or ten degrees.

Now have the person sit on the edge of the table and you perform the same manoeuvre, as illustrated. When the ankle

and leg are pulled outward and away from the midline, normally the hips will remain on the table in the same position; but when there is muscle tension or contracture, the hip on the affected side will rise off the table as the leg is pulled outward.

Now have the person lie face down on his stomach, as illustrated. Flex one knee so that one foot is up in the air, then gently move the leg outward from the level of the knee.

If there is muscle contraction and tension, the buttock on the side of the leg under examination will rise, or there may be some pain on that side. Watch for rolling of the pelvis away from the abnormal side during bending of the hip or elevation of the buttock.

TESTS FOR SCIATICA

The large sciatic nerve emerges from the spinal canal, most of its fibres coursing through the part of the iliopsoas called the psoas major, and goes on to innervate the muscles and skin of the legs and feet. If there is pressure and tension at the point of emergence, then sciatica results. There are specific tests to determine the presence of sciatica.

Flip Test

Have the person being tested sit up on a table with his legs dangling. He should hold his back as erect as possible with arms hanging at his sides. Hold your open palm on the thigh of one leg near the knee. Placing the other hand under his heel cord of the same leg, gradually straighten the leg. There should be no resistance or complaints until the leg is straight out. Continue gently extending the leg upward past this point. If the subject tends to flip over backward or braces himself with his hands to keep from falling on the table, sciatic nerve tension is indicated. When the flip test is positive, the patient will also be unable to sit erect on

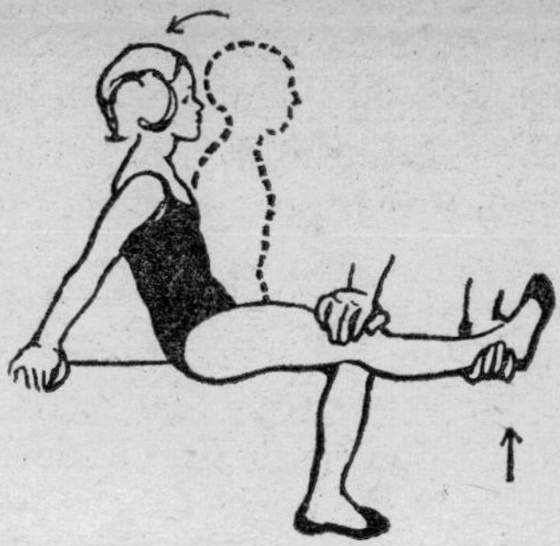

the table with both knees fully extended. Test both legs separately.

Then test both legs together, having subject place both legs straight out in front of him while sitting in place on table. Bending of the knees and curving of the back indicate sciatic tension.

Buckling Test

Have the person lie on his back on a table, with his legs on the table. Place one hand under the heel of one foot, and gradually elevate his leg above the table with the knee in full extension, as shown in the illustration. If the person's knee automatically buckles and his leg bends when it is raised about halfway, sciatic nerve tension is present. Test both legs.

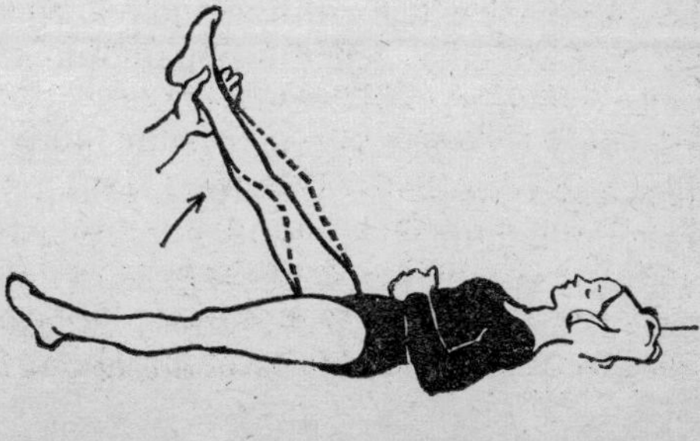

Interpreting the Muscle Tension Tests

If any of the results from these tests are positive, it means something is wrong. It means you have muscle contracture and tension. It means that certain of your muscles are pulling in an abnormal way on other muscles and on your joints. If you had a low score on the self-testing quiz at the end of Chapter I, but you did have trouble with these tests, then the chances are very strong that there is hidden muscle imbalance and that you eventually will develop some of the symptoms on the quiz. To correct the muscle problem that you have, and to prevent more symptoms from developing, you must now begin to stretch and relax your muscles by doing the exercises explained in the next chapter.

3: TEST FOR GENERAL PHYSICAL FITNESS

As long as you are doing tests, you might like to do some tests for general physical fitness to see how you stack up in that department also. People generally do not do as well as they think they will. The late President John F. Kennedy publicly commented on the fact that the rejection rate of draftees due to physical unfitness was so high that in order to get two soldiers, the Army had to call up seven men.

President Lyndon B. Johnson allocated millions of dollars for the rehabilitation of inductees rejected as physically unfit. This money would, perhaps, have been better spent for the training of medical and physical educators so that they could recognize and correct the various forms of physical unfitness by early categorization and management of the thirty per cent of the general population with muscle imbalance.

As I have said before, exercises taught in gymnastic and callisthenic classes are decidedly wrong for thirty per cent of the general population. Most of today's physical-fitness programmes are essentially for the benefit of the physically fit, and in actual fact are directed towards increasing athletic prowess rather than sustaining basic fitness.

What is physical fitness?

There is no universally accepted definition of physical fitness and no universally accepted method for measuring it. Some experts consider mobility the key, others, muscle strength, others talk about oxygen uptake by the body, and others use endurance or coordination as their measuring key.

Dr Kenneth H. Cooper, author of *Aerobics*, (M. Evans and Co, New York) the well-known book on the US Air Force exercise programme, measures fitness by the ability to jog or run at least one and a half miles in twelve minutes (or a little less for those who are over thirty-five or female).

The President's Council on Physical Fitness determines fitness by how many of a varied series of exercises a person can do.

The American Heart Association says fitness measurements should include the ability to pass the Masters Two-Step Exercise Test. This involves measuring the heart rate before and after the patient steps a certain number of times per minute, up and down two stationary nine-inch steps in the doctor's office.

In Sweden, great emphasis is put on physical fitness. Particular attention is being paid to the maximum oxygen intake as determined with the use of a bicycle ergometer. This was used quite successfully by the Swedish Olympic ski team, which won the Gold Medal team honours in the Olympics. If the oxygen intake is found to be decreased, it is known that there is something wrong. Other similar tests are based on the use of treadmills.

I will accept the results of any of these systems for measuring fitness. But I must point out that in none of these programmes can an orthophysically fit state be reached *until* the muscles are in good tone and balance. It is impossible for a person to achieve a trim figure, flexibility in movement, good posture, good physical fitness, and general good health unless his muscles are functioning correctly.

Take the following physical fitness test devised by Dr Hans Kraus of Columbia-Presbyterian Medical Center in

New York. This test indicates the minimum standard of fitness needed for an ordinary, non-athletic but healthy life. It has been used with school children by the thousand, and young people should be able to pass it with flying colours.

Remember, though, that this is a gauge of *minimum fitness*. If you pass, but barely, you are only just over the borderline of physical requirement for healthy daily living.

1. Lie flat on your back on the floor with your hands behind your neck; ask someone to hold your feet down and then roll up into a sitting position.

2. Again lying down with hands behind your neck, bend your knees and place your feet on the floor as close as you comfortably can to your body. Now roll up into a sitting position.

3. In the same position, raise both legs off the floor to a thirty degree angle and hold them straight out for ten seconds.

4. Lie on your stomach with a thick pillow under your hips. Clasp your hands behind your head and, while someone holds your feet in place, raise your head and chest off the floor and hold them up for ten seconds.

5. In the same position, lying face down, have someone hold your upper back down by pushing on your shoulder blades while you raise your legs off the floor and hold them up for ten seconds.

If you cannot pass this test of minimum fitness, you would do well to spend some *regular* time on the general conditioning exercises in the next chapter. Even though you may not yet have experienced a significant number of the previously described symptoms of muscle imbalance, if you are not minimally fit there is a good chance that you have a still-hidden muscle imbalance that will very probably show up to discomfort you some day.

CHAPTER FOUR

The Ortho-exercise Programme

These exercises are unique. They are ortho-exercises, specifically designed by orthopaedic specialists to work with the natural mechanics of your body. Some of them may seem familiar; perhaps they resemble exercises you have read about in some book or magazine article. But these are not the same as any other exercises. Even though several of them may resemble other exercises you are familiar with, there are important differences *in the way you do them*. This is because this is the only exercise system based on body mechanics.

You won't know just how much these ortho-exercises can help you until you start doing them. You will not get results by doing just any exercises in any way. To correct a medical problem, exercises must be tailored to your particular needs. In fact, as we have seen, the wrong exercises can make a condition of muscle imbalance worse instead of better, by causing the already overdeveloped muscles to develop even more, and by neglecting the muscles that really need exercise and loosening up. Thus, aches and pains can increase while weak muscles become weaker because mechanically incorrect strains are being put upon them.

The exercises in this book have been specially planned to use the principles of body mechanics for strengthening the muscles that should be strengthened to bring your entire body into proper balance.

Even after a few days on this exercise programme, you will begin to feel the difference. After several weeks you will start to notice the difference in your general health, in your walk and, most of all, in your hour-to-hour comfort.

Who Needs Ortho-exercises? Different individuals have different goals in their search for physical fitness. Those of the seventy per cent group are already fit for general conditioning exercises (pages 72–82). They will do well in any sport or activity. Still, they may seek an even greater degree of dexterity, proficiency, prowess, agility and accomplishment. I would refer this group to the many books written on the subject. Yet practically all physical-fitness programmes are directed to this group at the expense of the thirty percent-ers who require a physical fitness programme geared to their postural deficits and muscle imbalance.

I know a ten-year-old boy who presents a typical thirty per-cent-er's history. He spends most of his after-school time sitting on the sofa watching television, crayoning, or sitting at his desk, squirming about and trying to concentrate on his homework. Still, this child knows practically every member of the national baseball teams, their percentages at bat, whether they bat left or right, their standing with the team, etc. He owns a ball, bat and glove. Why isn't he playing ball with the neighbourhood kids? He has tried from time to time. When selected to play with a team, his performance at running, catching the ball or at bat is so poor that he declines to play again. When he screws up his courage to try for the team again, he is at the end of the list of candidates and is usually not selected. This child needs ortho-exercises. Any other physical fitness programme would only frustrate or harm him.

The Physical Examination. Regardless of your reason for doing physical fitness exercises, you should always begin with a complete physical examination by your physician.

In the young person under sixteen years of age the paediatrician will check for defects in the cardiovascular or pulmonary systems or any other pathology that could make exercise a dangerous activity for him.

The adult under thirty-five years of age should still get clearance from his physician, and his physician's assurance that he has no exercise limitations.

The physician may add an electrocardiograph to the physical examination of the adult between thirty-five and sixty-five. The doctor will check to make sure you do not have coronary heart disease or angina, any disease of the heart valves such as an old rheumatic fever, high blood pressure, uncontrolled sugar diabetes or excessive obesity. He will also check for kidney disease, anaemia, lung disease and arthritis.

The presence of any of these disorders does not necessarily mean that you cannot exercise. Each activity you engage in should be guided by medical judgement.

Do You Know How to Exercise? When I ask a patient, 'Do you know how to exercise?' he always laughs and answers, 'Of course.' When I show him how wrong he is, he is astonished. But we have all, almost certainly, learned the *wrong* way to exercise in all of our gymnastic and callisthenic classes.

There is really nothing difficult or extraordinary about the right way to do ortho-exercises. The catch is, you really have to *do* them. You are exercising *muscles*, not just improving arms or legs or chins. While you are doing them, you must think about the *muscles* that are involved. Feel them stretching, becoming looser and more flexible, now working *for* you instead of stopping your natural movements and grace.

Before you begin to exercise, put on some garment that will not impede your movement in any direction. A leotard is ideal; underwear, old slacks, or pyjamas – whatever you are comfortable in and that will not bind or pull at you in any way – not even at the cuff, the elbow or the collar. It will be all you can do to get your muscles moving again; you shouldn't have to struggle against a tight armhole too. Take your shoes off. You should have bare feet, so that they won't slip on the floor.

Now find a part of the floor on which you have enough space to sit with your legs spread out to the sides and to stretch your arms in every direction without hitting a lamp or a bed. You'll also need a wall at one side, or something rigid to put your feet up against. You'll see what other special space or equipment you need as you get into the descriptions of the actual exercises.

Next, try to set aside some time when nobody is going to interrupt you. If this is not possible, it's better to exercise with interruptions than not to exercise at all.

I always ask my patients to set aside twenty minutes to half an hour, twice a day, if possible, to see real results. After all, I tell them, years of neglect, years of letting your muscles go unsupervised, cannot be atoned for in a few minutes of lackadaisical stretching every now and then. It is not easy work – but the results are worth it. Imagine the freedom from pain and strain that will be yours when you have retrained your muscles to do their jobs properly. And, if you think you do not have enough time, realize how much easier it is to find a half-hour, even in a busy day, than it will be to get any work done at all when you are racked with the pain that comes when muscle imbalance continues uncorrected on its damaging course.

In reading over the exercise, you might be discouraged by the fact that you are asked to do some of them fifty, a hundred or even more times during one session. It sounds like an awful lot of work – too hard even to start, and even worse to continue. But don't be discouraged. You will be pleasantly surprised to find yourself in agreement with many of my patients. They have discovered that it is far easier to repeat one simple exercise fifty times than to go through a series of five exercises to be done ten times each.

The reason is logical. These, you must remember, are exercises designed to train muscles. When you do them correctly, your muscles are responding the way they are supposed to. They are getting longer, stronger, more flexible – and easier for you to use. The rapid repetition of a simple series of movements trains your muscles. Thus, you are likely to find that this exercise programme encourages you to continue on it, by making it easier and easier for you as you go along. Even when other more difficult stretches and bends are introduced, your muscles will have been prepared by the exercising that you did previously.

You may discover, when you start exercising, that in the morning you can hardly 'move a muscle' without feeling an ache or pain. This early morning stiffness is natural, and particularly noticeable during the first weeks or even months on an exercise programme. Remember, the purpose of these exercises is to stretch and limber up your muscles. And in the morning, after seven or eight hours in which most of your muscles have done very little work, they are bound to be stiff.

So if you find you can hardly get through your exercise schedule in the morning, stick to it anyway! This will get your muscles in condition for the activity they will be doing in the course of the day. Just go through your programme and move as much as possible; you'll probably find that after a few minutes your muscles have stretched noticeably, and exercising becomes more comfortable. Or you might try dividing your morning exercise session in half: do a brief run-through of a few routines for about ten minutes and then, an hour or two later, come back for fifteen or twenty minutes more. By this second inning you should be stretched enough for a vigorous and conscientious session. And you'll probably find that a morning exercise period leaves you, not tired out, but refreshed and 'rarin' to go'.

Commencement of the Active Exercise Programme. The best order of exercises to correct muscle imbalance is not always obvious. For example, the presence of a tight, tense or shortened iliopsoas will effect a continuing chain of secondary muscle contractures in such muscles as the adductor longus, tensor fascia lata, rectus femoris and quadratus lumborum – the technical names for the muscles in your buttocks, hips and legs. In any ortho-exercise programme, the initiator of the entire cycle is left for the end. The adaptations or adjustments are corrected first and the primary culprit – most often the iliopsoas – is corrected afterwards. It is particularly damaging for someone at the beginning of an orthotherapy programme to do such traditional exercises as press-ups. They only cause further muscular adaptations to the iliopsoas, and can cause more harm. Such exercises can only be undertaken when the adaptative rearrangement of muscles has been returned to normal and the full stretch of the iliopsoas has been completed.

How to Use the Exercises in This Chapter. In this chapter you will learn the twenty-one basic exercises for restoring muscle balance. These exercises have been designed for specific muscles that are most often in need of loosening and strengthening.

There is also a second section of conditioning exercises for those who do not have any muscle problems but would like to keep their present flexibility of movement. These conditioning exercises can also be done by those who do have muscle imbalance, but only *after* they have been on an ortho-exercise programme, when they feel that most of their symptoms have been alleviated and they have passed the muscle tension tests. These general conditioning exercises will help prevent them from sliding back into their former condition.

How Long Should a Programme of Ortho-exercises Be Continued? After a few weeks you may already notice that you bend better, stretch better and reach higher, that your once stiff joint will now move in all normal directions without any difficulty and that your other symptoms are starting to decrease. You will be tempted to stop. But don't stop yet. You can reduce the exercise period, but you should not stop completely. If you stop exercising, you give the muscles a chance to contract again. After several months of gradually reduced exercises, if none of the pain or discomfort has returned, you can then stop the corrective exercises and just keep fit with the general conditioning exercises. However, every few months you should again take the tests in Chapter III to make sure that the condition is not returning.

Always check first with your doctor, but to help you to

determine which exercises you should do to help your specific problem, see Chapter V (Designing Your Own Ortho-exercise Programme.) When you are ready to start exercising, first read the directions and slowly go over each movement described. Check to see that you are in the proper position by referring to the illustrations. You may wish to have someone read the directions to you as you act out what he reads. Then, go to work! It will be worth it, I promise you.

THE TWENTY-ONE ORTHO-EXERCISES FOR RESTORING MUSCLE BALANCE

1. Arm Circling

Stand with your feet slightly separated. Bend forward with your arms and head hanging down loosely. Bring your arms forward, up and back in a circle (like a windmill). This exercise can be done with one arm at a time or both together, whichever is more comfortable for you. Make from fifty to 300 continuous circles with your arms at least once a day. This exercise is good for loosening up the muscles of the shoulders, shoulder blades and upper back. If your shoulders and/or shoulder-blade musculature is very stiff, you will be able to make only small circles at first. You are the person who should make at least 300 circles a day. As your range of motion increases you will be able to make

larger and larger circles and will be able to cut down their number. Aim for large, free-swinging circles with strong participation of the shoulder-blade muscles. When you experience cramping and pain in the neck and shoulders this is the first exercise to do. Repeat it for about two minutes every half-hour during the day, whenever it's convenient. As pain and discomfort subside, you can gradually decrease the number of times.

2. Shoulder Shrugging

Stand up straight, with your feet slightly separated, and let your arms hang loosely at your sides. Shrug your shoulders in a circle by bringing them up as high as possible, then as far backward as possible, pulling the shoulder blades strongly together at the back, then relaxing the shoulders into the starting position. Many people tend to thrust their head forward while doing this exercise; this should be avoided.

Hold your head up, and your neck as straight as you can. Repeat the shrugging motion twenty to fifty times. This exercise will loosen up the muscles of the shoulders and neck. If you are having pain or stiffness in these areas you can repeat this exercise as many as ten times a day. Otherwise, once or twice a day will suffice.

3. Shoulder Blade, Upper and Middle Back Strengthening

Stand with your feet wide apart and body bent forward at the waist. Try to bend forward in a relaxed way by putting your head down and letting its weight pull your torso over and down towards your feet. Clasp your hands behind you at the waist. Now, remain bent at the waist and lift your torso by raising your head and arching your back while pulling

your shoulder blades sharply together. Hold this position for a fast count of ten. Then relax forward again. Repeat ten to twenty times. Whenever your upper back, neck and shoulders feel cramped and painful you should repeat this exercise three or four times a day. Otherwise, once or twice a day will suffice. This exercise will help correct any exaggerated forward or backward curvature of the spine.

4. Shoulder and Shoulder-Blade Strengthening Against Resistance

This exercise requires the assistance of another person. Lie face down on the floor with a pillow under the abdomen and another under the ankles to help you relax. Have the other person kneel at your left side, place his right hand at the base of your spine and his left hand on your right shoulder blade. (A) Now holding both arms along your sides, raise your right shoulder as far as you can off the floor while turning your head to the right. The pressure of the other

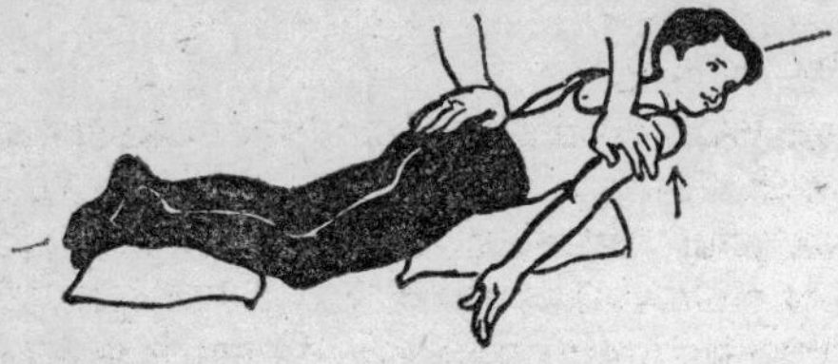

person's body weight against your shoulder will make this quite difficult at first. Don't despair. Raise the right shoulder and slowly lower it ten times. Then shift to the other side and raise the other shoulder blade ten times. (Your aide will have to shift so that he is kneeling on your right side, placing his left hand at the base of your spine and his right hand on your left shoulder blade.) (B) Now shift your arm position by holding your upper arms straight out to the sides; bending your arms at the elbow; and pointing your lower arm, hand and fingers up over your head. Have your aide resume his position on your left side,

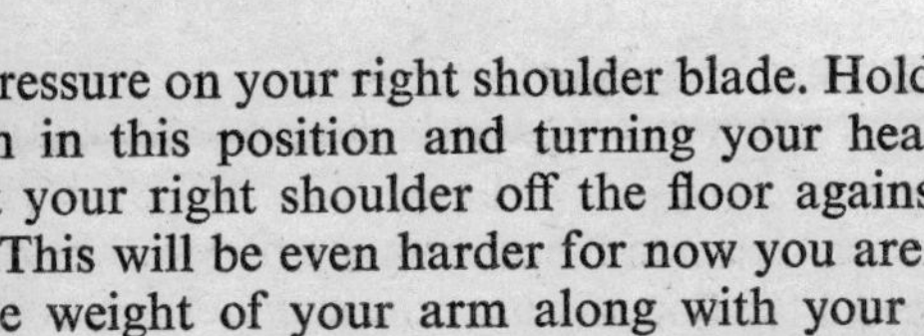

placing pressure on your right shoulder blade. Holding your right arm in this position and turning your head to the right, lift your right shoulder off the floor against his resistance. This will be even harder for now you are carrying the whole weight of your arm along with your shoulder against the resistance. Repeat ten times. Now follow suit with the left shoulder blade and flexed left arm. Remember, the more difficult an exercise is for you, the more beneficial it is. If you can't find someone to aid you in this exercise, do shoulder lifts A and B without resistance. It will still do some good. Do this exercise sequence two to five times at least once a day.

5. Shoulder Rotation

Lie on your back on the floor with your knees up and your feet flat on the floor, arms along your sides. Place your feet as close to your buttocks as is comfortable for you and about six or eight inches apart. Hold a weight of two to five pounds in each hand (tinned goods work well). Raise your arms up into the air holding your elbows straight and then, still keeping your elbows straight, stretch them over your head, and try to reach back and touch the floor. At first, you

may not be able to keep your arms straight and still touch the floor; this indicates that the adductors and internal rotators (the muscles that pull and turn the arm in at the shoulder) of the shoulder joint are tight. When you are able to do this you will see how much these muscles have loosened up. Repeat ten to fifty times at least once a day.

6. Upper Back Muscle Strengthener

Kneel on the floor and bend forward from the waist until your forehead touches the floor. You will probably want to protect your forehead by putting a pillow on this spot on the floor. Keep your thighs perpendicular to the floor, and clasp your hands behind your back at the waist. Now lift your head and raise your torso so that it is parallel to the floor by arching your back and pulling your shoulder blades strongly together. Your hips will rock back somewhat, but try to

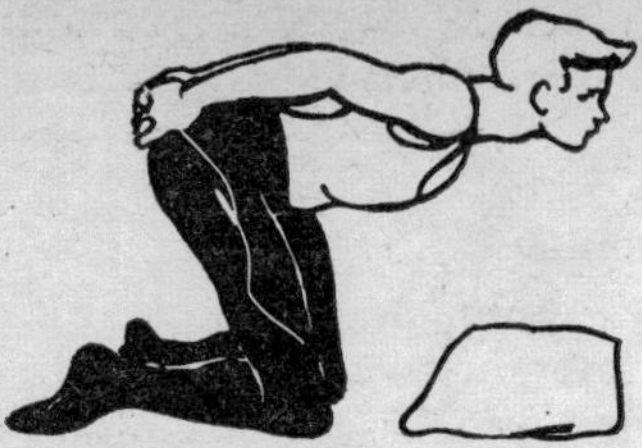

control this as much as you can. Hold this position for from five to ten seconds, then relax forward and rest your forehead on the pillow for about five seconds. Repeat five to ten times at least once a day. This is a difficult but most rewarding exercise.

7. Low Back Stretch

Sit on the floor and move your left leg as far as possible towards your left side. Flex the foot so your toes are pointing straight up at the ceiling. Moving your right leg to your right side, bend the right knee, and bring the right heel in close to the crotch, keeping the right knee on the floor and holding your left hand in the small of your back. (A) Sitting as erectly as possible, twist your torso to the left until you are facing your outstretched left leg. Now reach out your right hand and try to touch your left toes, bending your torso from the hips. Bounce your torso in this position, reaching your arm towards your toes. Your whole upper body, head and right arm should be stretching towards that foot. Bounce hard and rapidly 100 times. You may not

be able to do this 100 times at first, but keep trying. Now shift position and bounce your left hand to your right foot 100 times. Repeat with each leg once more if you can and do this exercise at least once a day. (B) For added effect an assistant may get behind you on the floor and place one of

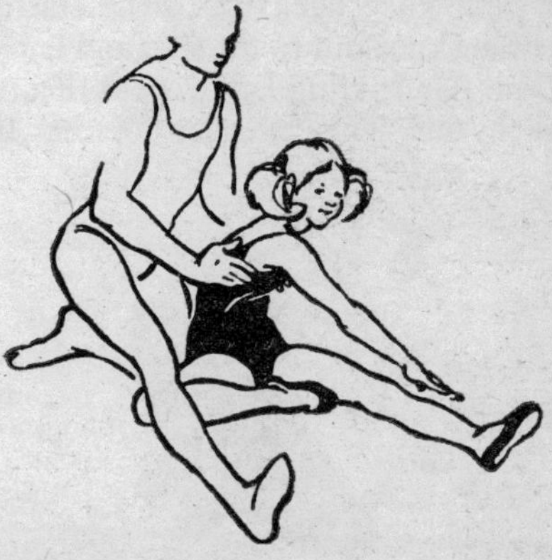

his legs over your bent knee to stabilize you while kneeling on his other leg. He can also grip your torso, while bracing you in that position, and help you turn your torso towards your outstretched leg. This helps you to get more of a back twist than you can achieve by yourself.

8. Leg Roll

Lie on your back on the floor with your head resting on a pillow and your hands crossed on your abdomen. (Your upper body should be as relaxed as possible.) Bend your knees up, placing your feet flat on the floor as close to your buttocks as is comfortable for you and about twelve inches apart. Moving the bent leg from the hip joint, roll one knee out to the side as far as possible. Ultimately, you should almost be able to rest the bent leg on its side on the floor.

Now roll the bent knee back up to the starting position. Then roll the other knee out to the side and back up into the starting position. Alternating legs, repeat this ten to twenty times with each one. This exercise loosens the hip-joint

structures. If it is very difficult for you to do, or if you have pain or cramping in the hip, do this exercise four or five times a day at first, until the pain lessens and muscles loosen up.

9. Knee-Chest Stretch

Lie on your back on the floor with a pillow under your head and your arms resting alongside your body. Bend your knees up with your feet flat on the floor as close to your buttocks as is comfortable for you. Keep your feet about twelve inches apart. (A) Clasp your right knee with your

right hand and pull your knee as close to your chest as you can. Rock your knee towards your chest twenty to fifty times. Then repeat, holding the left knee with the left hand. (B) Now, from the same starting position, pull both knees up to the chest at the same time, using both hands. Keep the knees close together and as close to your chest as you can

for the count of ten. Keep your shoulders flat on the floor throughout. Repeat this exercise for from one to three minutes a day.

10. Long Body Stretch

Kneel on the floor with your knees about six to eight inches apart. Bend your torso forward from the waist, stretching your arms out over your head and holding your elbows straight so that your forehead and your lower arms and hands are resting on the floor. Your thighs should be held

perpendicular to the floor and your whole upper body should slope down from your hips to the floor. Being sure to keep your thighs perpendicular to the floor, press your chest as close to the floor as you can. Press down for a fast count of ten, then relax the chest but stay down in the starting position for a count of five. Repeat as many times as you can in three minutes. This exercise utilizes the force of the body's weight and line of gravity to stretch the hip joints, the entire spine, the adductors and the internal rotators of the shoulder joints.

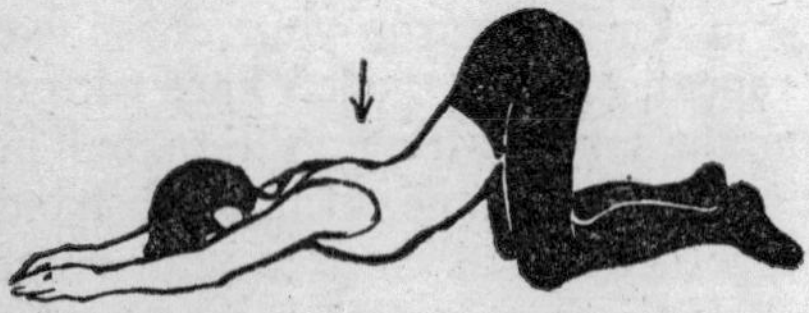

11. Flat-Back Position

Lie on the floor on your back, resting your head and upper shoulders on a pillow. Cross your hands on your abdomen. Bend your knees and rest your feet flat on the floor about twelve inches apart and as close to your buttocks as is

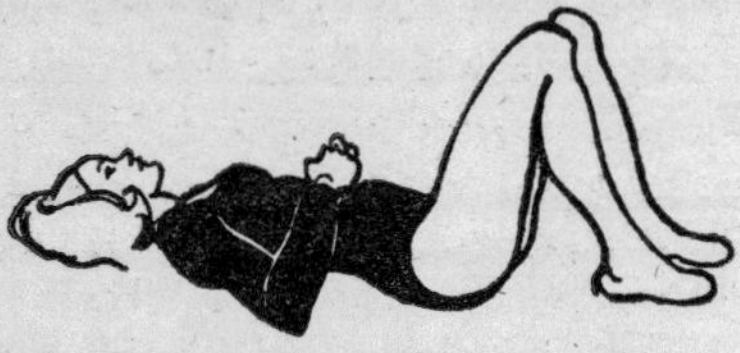

comfortable for you. Your head, shoulders and legs should be as relaxed as possible. Now tighten the buttock muscles

and pull the abdomen in at the same time. You will feel the curve of your back flatten against the floor. Hold for a count of ten. Relax for a count of five. Repeat for one to two minutes at least once a day. This exercise stretches the lower back muscles and tones up the seat muscles.

12. Chair Droop

Sit forward on a hard chair. Keep your knees slightly apart, feet flat on the floor and relax your arms at your sides.

Now let your torso go limp like a rag doll and allow the weight of your head and arms to pull your torso down between your legs as your head and arms fall forward. If you are loose enough, your head will fall between your knees and your hands will flop over your feet. Now tighten your abdominal muscles and slowly roll your spine back up until your body is once again in sitting position. Repeat six times, once a day. This exercise will stretch and loosen lower back muscles.

13. Knee Bends

Find a stable table or chair that is about as tall as your waist. Stand facing it with your hands resting on it for balance. Leave as much room as is comfortable between you and this support – you shouldn't have to lean forward to reach it. Now squat down, bending at your knees and hips, as deeply as you can. Hold your spine as straight (perpendicular to the floor) as you can. Squat for a count of ten, then slowly stand up again, using your leg muscles as much

as possible and trying not to pull yourself up by bracing your hands on the supporting table or chair. Repeat six times, once a day. This simple exercise will strengthen foot, leg, thigh and lower back muscles. It also helps you to practise good body alignment.

14. Hamstring Stretch

Stand on one leg in front of a table or bench that is about as high as your hips. Stretch the other leg straight out, resting your heel on the table with your toes pointing up. (A) Bend forward from the hip, reaching towards your out-

stretched foot with both hands, and bounce your torso hard, trying to reach farther towards your foot each time. Bounce 100 to 200 times, then switch to the other leg and bounce 100 to 200 times. (B) If you are round-shouldered and/or have a rounded upper back, assume the same starting position, but bounce the torso towards the outstretched leg while

holding your spine straight, shoulders back and head up in a 'good posture' position, arms at your sides. Bounce 100 to 200 times with each leg.

15. Hamstring-Gastroc-Soleus Stretch

Sit on the floor with both legs outstretched in front of you, feet propped up flat against the wall. (A) Bend forward from the hip as far as possible, stretching your arms out towards your toes. Bounce forward hard, trying to reach

your toes 100 to 200 times. This stretches the tight 'hamstring' muscle in the back of your thigh and the gastroc-soleus muscle in the calf. (B) If you are round-shouldered, and/or have a rounded upper back, assume the same starting position, but bounce the torso forward while holding your spine straight, shoulders back and head up in a 'good

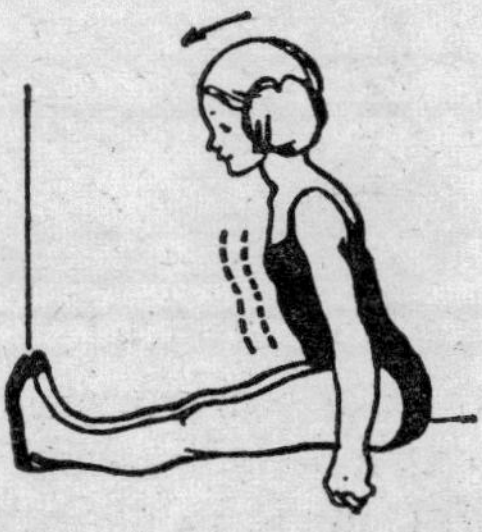

posture' position, arms at your sides. Bounce 100 to 200 times. This will help to straighten and strengthen your upper back while stretching the thigh and calf muscles.

16. Heel and Calf Muscle (Gastroc-Soleus) Stretch

Stand facing a wall or chair at arm's length away from it. Put both hands flat on the wall, or hold the top of the chair with both hands, for support. Keep your feet flat on the floor with your toes pointing straight ahead. Now move your pelvis forward in short, jerky movements, as though you were doing bumps in a burlesque. You will feel the pull in your calf muscle. Continue doing these forward pelvic movements 100 to 200 times. For a more powerful stretch of the calf muscles, place a block two to three inches high under the forefeet.

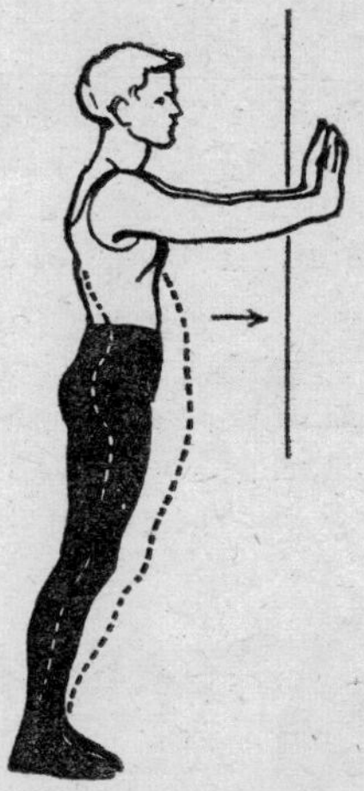

17. Inner Thigh (Adductor) Stretch

Sit on the floor with your legs stretched out as far apart as possible. (A) Bend the body forward between your legs, reaching forward with the arms and trying to stretch your body as close to the floor as possible. If you can't seem to get your body to move forward, you may want to give yourself a push by bracing your arms behind you on the floor and pushing the body forward from the rear. Or you may

want to grab both knees and pull your body forward. As your muscles loosen and your range of motion increases you will be able to stretch your legs farther apart and bend your body farther forward. Bounce your body forward (be careful to hold the knees straight) 100 to 200 times. (B) If you are round-shouldered and/or have a rounded upper back, assume the same starting position, but lean the body forward

between the legs holding your spine straight, shoulders back and head up in a 'good posture' position. When you first start this exercise you will almost certainly have to attain the forward bounce of the body by bracing your hands on the floor behind you and pushing forward from the rear. As you limber up, you will be able to bounce forward with a straight back, arms hanging loosely at your sides. Bounce forward 100 to 200 times. (C) If you find this exercise tremendously difficult, loosen up first by lying on your back with your knees bent up close to your chest. Grabbing the right knee with the right hand, move it out to the right side and move it in large circles (so that you feel your thigh

rotating in the hip socket) first clockwise and then anti-clockwise, twenty times each way. Now grab the left knee with the left hand and do the same. Repeat cycle until you have spent from two to five minutes on each leg, at least once a day. When you can do this circling exercise without discomfort, go back to the stretching exercise and master it.

18. Hip and Thigh Connector Stretch (Tensor Fascia Lata)

Stand at arm's length from a wall or chair with your side to it. Place the hand closest to the wall or chair on it for support. (You should not have to stretch to reach this support. Stand as close as you need to reach it easily.) Bounce the hip facing the wall sharply towards the wall, so that your whole pelvis is jerked to the side. Repeat this sideways bounce twenty to fifty times. Turn round and repeat on the other side. You will feel a stretching sensation at the outer side of the hip closest to the wall or chair. This exercise stretches the tight hip and thigh connector muscle, increases the range of motion, thus facilitating correction of hip and thigh disalignment.

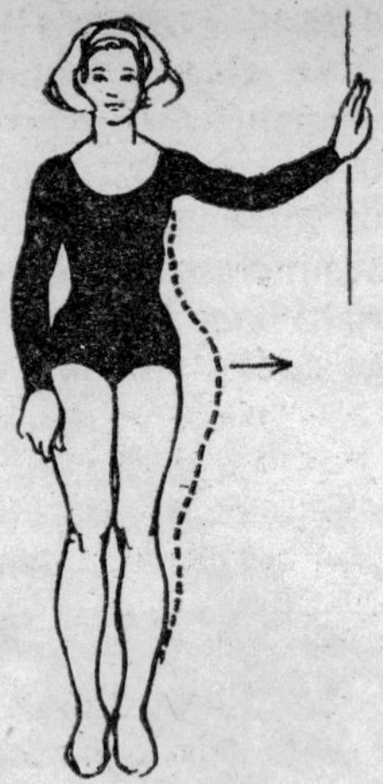

19. Fencer's (Iliopsoas) Stretch

Get into a fencer's thrust position, as illustrated. Place your right foot forward, bending your knees and stretching your leg as far in front of you as you can. Turn your right foot in slightly. Stretch your left leg out straight behind you so that your foot is straight ahead, braced against the floor

with the heel off the floor. Look behind you and check to make sure your left foot is not turned in. Hold your torso erect and bend back from the waist slightly. Bounce your torso backward until a pull is felt in the groin. (You may want to put your left hand on your left hip and rest your right hand on your right thigh to help balance yourself.) Do fifty bounces on your left leg, then switch positions so your right leg is back and do fifty bounces. The chronic sloucher will feel a pull in the muscles in his abdomen and chest. This excercise is a critical one as it stretches the iliopsoas, and aids body flexibility and alignment. Repeat this exercise as many times in the day as you have time and strength for.

20. Manual Iliopsoas Stretch

This exercise cannot be done without the help of another person, so I usually only insist on it for children who have parents to help them out. The exercise is good for people of all ages, so if you have a cooperative husband or friend, try it. Lie on your back on a table that is high enough for your legs to dangle freely from the knee. First, bring your right knee up to the outer side of your chest, as far as you can, and hold it there with both hands. The small of your back should be flat against the table. Now have your aide press

down on your left leg to flatten it against the table. If your left leg doesn't flatten against the table without pressure from an aide, it means the iliopsoas muscle on that side is too tense or shortened. Even with pressure applied you may not be able to flatten the left leg against the table at first. Have your aide press against your left leg in short downward thrusts twenty to fifty times. Switch legs, holding your left knee up against your chest and have your aide press down on the right leg twenty to fifty times. Repeat this exercise as many times during the day as you have time and strength for.

21. Front Thigh (Rectus Femoris) Stretch

When you tried to do exercise 20, if the bent knee of the loose leg tended to extend or straighten up as the thigh was pressed down to touch the table, you have a tight rectus femoris, or front thigh, muscle. The following exercise is particularly good at correcting that problem. Kneel on the floor with a small pillow under your ankles. Clasp your hands behind your head and, holding your whole body from the knees

up in as straight a line as you can, lean back from the knees. Holding this position (your body is forming a seventy-five-degree angle with the floor) bring your pelvis forward as far as you can and then backward as far as you can, slowly and carefully, without moving your feet or knees. Repeat this backward and forward slow wiggle twenty times, at least once a day. Admittedly this will be difficult at first, but you can master it if you persevere.

THE SIXTEEN GENERAL CONDITIONING EXERCISES

The following exercises are recommended for those who are not in need of special or corrective exercises, or for those who have undergone and completed a corrective ortho-exercise programme and are now ready for general exercises that will keep their muscles in equilibrium and good working condition. You should do these exercises only if you no longer have any signs of muscle imbalance. You should *not* do these exercises if you still have muscle imbalance and cannot yet pass the physical tests in Chapter III.

1. Jogging

Run on the spot for one to two minutes, easily and loosely. Be sure to come down on the ball of your foot, not on your toes. Don't leap high with each step or come down heavily. This exercise is a warm-up for the muscles and joints and prepares the heart and lungs for the exertion ahead.

2. Touching the Floor

Stand with your feet about two feet apart pointing as straight forward as is comfortable for you. Bend forward from the hip and try to touch the floor with both hands, keeping your legs straight. First bounce your torso and try

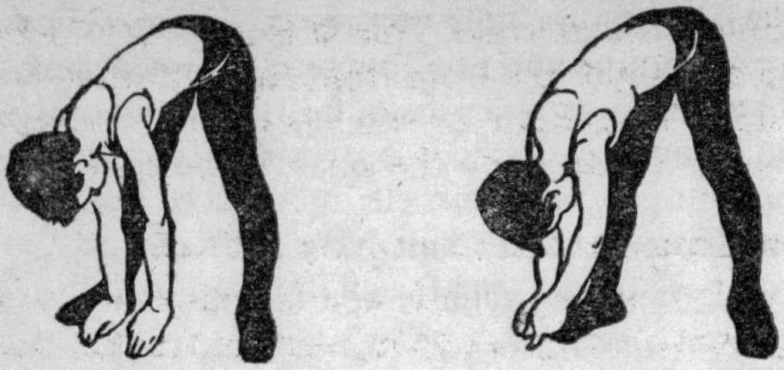

to touch the floor between your legs ten times; then shift again and try to touch your left foot with both hands ten times. Then touch your right foot ten times. This exercise loosens up the muscles and joints of the back, hip, shoulders, pelvis and legs.

3. Sideways Bend

Stand with your feet one and a half to two feet apart, hands hanging at your sides. Bend to the right, sliding the right hand as far down along the outside of the right leg as possible. Straighten up and repeat ten times. Repeat, bending to the left ten times. Many people will bend forward slightly

while doing this exercise. This is wrong. Try to keep your body in a straight line from front to back while bending to each side. This exercise stretches and strengthens the muscles and joints along the sides of the spine.

4. Front Shoulder- and Chest-Muscle Stretch

Stand with your feet slightly apart, arms raised to shoulder height, elbows bent, and your hands in front of your chest. Keeping your elbows bent, bring your upper arms back and tighten your shoulder blades to a 1–2 count. Straighten

your elbows and swing your whole arm straight back, keeping them at shoulder height, and bring the shoulder blades together to the count of 3 and 4. When you bring your whole arm back your thumbs will be pointing towards the ceiling. Repeat the entire 1–2–3–4 count five to ten times. This exercise stretches the chest muscles to improve posture. Check to make sure your head does not jut forward while doing this exercise.

5. 'Fall-out'

Stand with your hands on your hips and your feet together. Now simultaneously move your right foot out to the side, bending the right knee and placing full weight on the right foot, while bending the head and torso to the left side.

Now switch to the other side so that weight is on the left foot and the torso is turned to the right. Repeat on both sides five to ten times. This exercise improves balance and coordination.

6. Sit-ups

Lie on your back on the floor with your knees bent up, feet flat on the floor, slightly apart, and as close to your buttocks as is comfortable for you. Clasp your hands behind

your head. Do five to ten sit-ups in a curling fashion, bringing your body straight forward. Then do five sit-ups twisting your body to the left, and five twisting your body to the right. This exercise strengthens and tightens the abdominal muscles.

7. Side-to-Side Pull-ups

Lie on your back on the floor with your arms straight out to the sides. Bend your knees and place your feet flat on the floor as near your buttocks as is comfortable for you. Keeping your arms out to the side, pull your legs over to the right, knees still bent, and strongly up on your right side towards the chest. Return to starting position. Now pull the legs over

to the left and strongly up towards the chest. Continue, alternating sides, ten to thirty times in all. This exercise strengthens the abdominal muscles and hip muscles and stretches the lower back and spinal muscles.

8. Leg Swing

Lie on your back on the floor with arms straight out to the sides and your legs straight up in the air so they form a right angle with your body. Keeping your shoulders and arms flat on the floor, swing your legs over to the right as far as you can. This will require a twist of the torso to the right. Both legs are still together and still form a right angle with the body. Swing legs back up into starting position and repeat to the right ten times. Repeat to the left ten times. This exer-

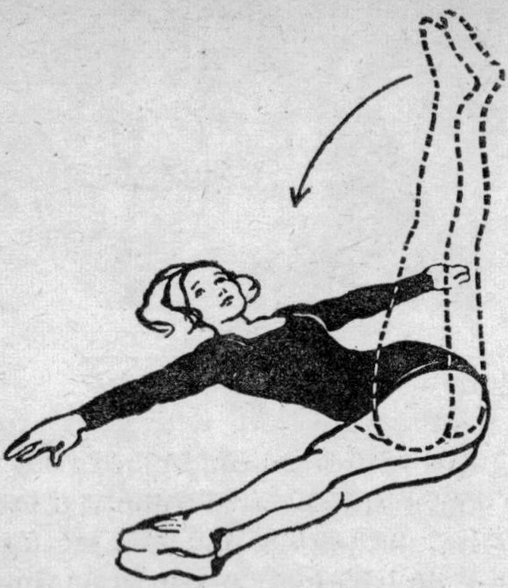

cise strengthens the abdominal and trunk muscles and stretches the intervertebral joints of the spinal column.

9. Pelvic Rotation

Lie on your back with your arms at your sides, knees bent, and feet flat on the floor as near your buttocks as is comfortable for you. Raise your hips, putting your weight on your shoulders, so that your body forms a straight line from your shoulders to your knees. From this position rotate your pelvis in a circular motion ten to twenty times, first to the left and then to the right. This exercise strengthens and loosens up the back, pelvis and hip structures.

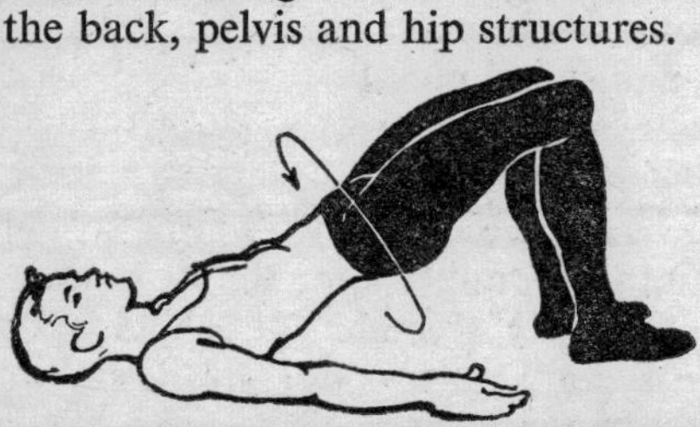

10. Inner Thigh Stretch

This is the same as ortho-exercise 17, part A. Sit on the floor with your legs stretched out, as far apart as possible.

Bend the body forward from the hip, reaching forward with the arms and trying to stretch your body as close to the floor between your legs as possible. If you are up to the conditioning exercises, you should be able to do this exercise easily without any push from your arms as in the ortho-exercise. This is one exercise you should never stop doing. It stretches the muscles on the inside of the thigh – a stretch which almost everyone needs, no matter how fit he is.

11. Clap and Jump

Stand with feet together, hands at your sides. Jump so that your feet are far apart and clap your hands over your head

at the same time. Jump and bring hands and feet back to starting position. Repeat twenty to fifty times. This exercise will increase coordination and loosen up muscles.

12. Rib-Cage Swing

Stand with feet from one and a half to two feet apart. Bend forward from hips, with your head and arms hanging down loosely. Straighten up and swing both arms as far up and backward as possible. Expand your chest as you throw your arms back. Then relax forward again until head and arms are hanging down loosely. Repeat ten to twenty times. This exercise will loosen up muscles and joints of the back, pelvis, hips, shoulders and rib cage.

13. Side-Kick

Lie on your left side with your left leg bent slightly at the hip and knee, so your body is braced and does not tilt either forward or backward. Raise your right leg straight up ten times, keeping the knee straight and toes pointed straight forward. Turn to the left side and repeat. You will feel the pull on the muscles at the outer side of the abdomen

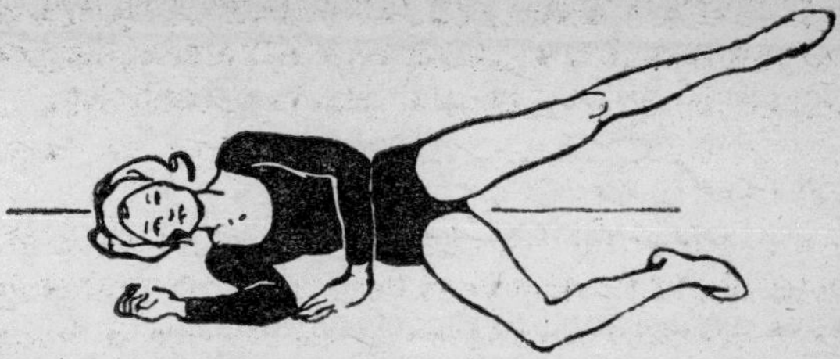

and hip on the side of the moving leg. This exercise strengthens these muscles.

14. Upper Torso Strengthener

Lie face down on the floor with a pillow under your pelvis, and your hands clasped behind your back at the waist. Raise your head, shoulders and upper body, bringing the

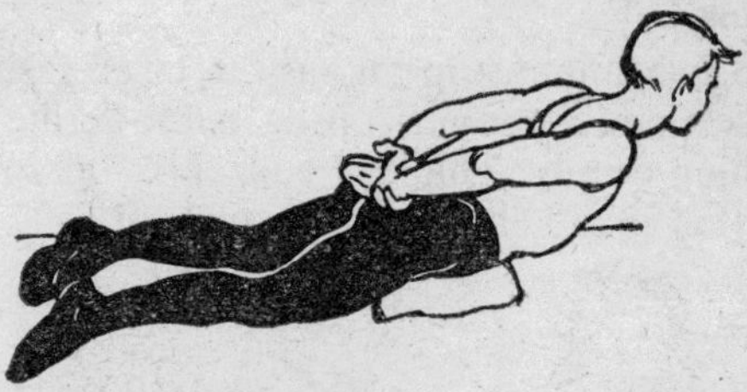

shoulder blades strongly together and arching the back. Slowly lower head and torso to starting position. Repeat ten to twenty times. This exercise will strengthen the back extensor muscles as well as the shoulder-blade adductor muscles.

15. Full Torso Stretch and Strengthener

Lie face down on the floor with a pillow under your pelvis and your arms out to the side, bent at the elbow so that your fingers are pointing over your head. The upper arms should be on a line with the shoulders. Keeping your knees as straight as possible, raise your right leg, your head, and your left arm and shoulder, all at once. Your weight is now on your right arm and left leg. Holding this lift, twist

your trunk as far round to the left as you can, further lifting your left arm and shoulder. Relax and return to starting position. Then repeat on other side by raising your left leg, head, and your right arm and shoulder, and turning your torso to the right. Repeat this, alternating opposite arms and legs, 100 to 500 times. This exercise will strengthen and build up the endurance of the muscles of the back, shoulder blades and hips as well as toning up the abdominals.

16. Tapering Off

Jog one to two minutes in an easy, relaxed fashion. Then slowly walk round the room, raising and lowering your arms and breathing deeply. This lets you taper off so that your body function gradually returns to normal.

CHAPTER FIVE

Designing Your Own Ortho-exercise Programme

In this chapter we show you how to use the exercises in Chapter IV for relieving your own aches and pains. There are also some special exercises described here, for relief of specific symptoms. You may find that most of your symptoms are completely eased by the exercises. Other symptoms may be helped somewhat, but will still need additional treatment. If, while on your exercise programme, you find your symptoms get worse or if new symptoms occur, see your doctor promptly.

When you have experienced relief of the painful symptoms that you are attempting to correct, continue exercising for several weeks longer. Then, gradually, you can taper off. Put yourself on a maintenance programme of exercising for shorter periods, perhaps fifteen minutes a day three or four times a week. If symptoms return, resume the more intensive exercise schedule. You can double-check on the extent of your progress by trying again the physical tests in Chapter III. Whenever you find that your muscles are tightening up again, or if any other discomforts return, start exercising again.

STIFF OR ACHING NECK AND HEADACHE

A sudden neck pain can be caused by injury, but the usual neck pains are chronic ones that occur intermittently over months or years. You should see your doctor if you have persistent neck pains, since they can be caused by bone destruction, vascular abnormalities or other complications;

but for the everyday variety of neck pains and headaches, the following do-it-yourself treatments will probably help.

Sleep with a small, soft pillow to avoid unusual angles or stress on the neck during the night.

Sleep on your back or on either side, not with your face down.

When you have an attack, apply heat to your neck for twenty minutes each evening or, if possible, twice a day. Use a heat lamp, heating pad, hot towel, or simply take a long, hot shower.

To relieve severe pain, a collar support that you can make from a towel is helpful. Fold the towel lengthwise four times and wrap it snugly about your neck. It should encircle your neck about one and a half times. Overlap the towel beneath your chin and anchor the end of the towel at the side of your neck with two safety pins. Or you may use a stocking stuffed with cotton or sponge rubber in the same way. These are more comfortable than rigid commercial collars. You can wear one of these when sleeping or whenever you want.

And do ortho-exercises 1, 2, 3, 4 and 5 as described in Chapter IV, followed by this special exercise: gently roll your head in a full circle, first completely round to the right, then to the left. Ortho-exercises 6, 10 and 19 will also help prevent recurrence.

Take aspirin, as needed.

If these measures do not give relief in five to ten days, see your doctor.

PAIN IN THE NECK AND SHOULDERS (SCAPULOCOSTAL SYNDROME)

Most neck and shoulder pain is a chain-reaction result of poor posture producing muscle tension. But in more extreme cases there may be painful tenderness in certain areas on the muscle, or severe muscle spasms. This is known as scapulocostal syndrome.

During an attack it is helpful to apply heat (from moist hot packs or a heating pad) to the neck and upper back for twenty to thirty minutes. This should be followed by massage and ortho-exercises 1–6, 10, 19 as described in Chapter IV. Repeat the exercises several times a day during periods of great discomfort, and at least twice a day at other times. If severe attacks continue despite the exercise programme, see your doctor. He can sometimes give relief of pain by applying firm pressure over a nerve trigger point, or he may inject procaine or hydrocortisone into the area or use an ethyl-chloride spray to stop the pain.

ROUND SHOULDERS

Round shoulders and slumped back are actually chain-reaction results of rigid hip muscles, which tilt the pelvis forward and force the person to curve the small of his back and hunch his shoulders in order to maintain balance while standing or sitting. This results in what we recognize as 'poor posture'.

Frequently, round shoulders are accompanied by exaggerated lumbar lordosis, or sway-back – too deep a curve in the lower back with the seat sticking out. Often the person has a flat-footed stance as well.

It is important to understand that it will take more than will-power to correct posture problems such as these. They are symptoms of muscle imbalance, and only strengthening by exercise of the leg and hip muscles will get at the root of the problem.

Start with fifteen or twenty minutes, twice a day, of ortho-exercises 1, 2, 3, 4, and 5 in Chapter IV, doing them slowly 100 times each. At weekly intervals, speed up your rate on these ortho-exercises and add ortho-exercises 6, 7, 10, 14, 15, 17, 19, 20 and 21. By the fourth week you should be able to see and feel the improvement.

LOW BACK PAIN (LUMBAGO)

This is the most common symptom of muscle imbalance. Millions of Americans have disabling back problems. The National Safety Council estimates that back problems each year cost the American economy well over one billion dollars in lost goods and services and Workmen's Compensation payments. Some medical authorities estimate that one out of every two Americans eventually has some type of lower back pain, causing him days of misery.

Dr Hans Kraus, consultant to President John F. Kennedy, who designed the minimum physical fitness test in the beginning of this book, told doctors at a medical convention in New York City that more than eighty per cent of the pain in the lower back is muscular in origin. In fact, he said, nearly all pain in the joints is usually due to muscle imbalance rather than to local defects of the bones or joints. He recommended, and I could not agree with him more, that every general medical examination include testing of the mobility and strength of the skeletal muscles in this area, and that as soon as muscle imbalance is detected, corrective exercises should be prescribed.

If you are one of the millions who suffer from low back pain, you can reduce strain on your back in all of these ways:

When you sit, use a hard chair with a straight back and put your spine up against it. Contour chairs like the ones used by typists, offer excellent support. If possible, use a footstool to raise one or both knees higher than your hips. During prolonged periods of sitting, cross your legs, to rest your back; and alternate your crossed legs occasionally.

When you stand, try to keep your lower back flat; tuck your buttock muscles in to straighten out the small of the back. When you must work in a standing position, put one foot up on a foot-rest to help relieve sway-back and lower back muscle tension. (A stool or even a brick or a child's large block can be used as a foot-rest.)

Never lie on your stomach to sleep. If you lie on your side, curl up with your legs bent at the knees and hips. Some people experience greater relief by placing a small pillow between the knees.

For driving, place a hard seat-and-back-rest combination over the seat of your car. Sit close enough to the steering wheel for your knees to remain bent, so they do not fully extend when you work the pedals.

It is of the utmost importance to lift heavy weights

properly. Bend your knees and use your leg muscles to provide the lifting force. Avoid sudden movements. Avoid lifting anything heavy above your head.

Avoid carrying unbalanced loads – use two tote or shopping bags with weight about equalized. Hold heavy objects close to your body, not at arm's length. Never carry anything heavier than you can manage with ease, and try not to move or lift heavy furniture. If you must lift or move heavy objects, whenever possible get some guidance from someone who knows the principles of leverage.

Don't overwork yourself. If you can, vary your working position by changing from one task to another before you feel fatigued. If you must work at a desk all day, get up and move around periodically.

If you experience an acute attack, try for immediate relief with aspirin (or a stronger prescribed medication), rest in bed, and warm baths. You may also find relief and support by taping or pelvic traction, hot packs, whirlpool baths, certain sedative or muscle relaxants or manipulation.

Always sleep on a firm mattress: or place a plywood board, three-quarters of an inch thick and the same length and width as your mattress, under it.

Check with your doctor; but in general don't do exercises that strain your lower back, such as backward or forward bends, touching the toes with the knees straight, press-ups or pull-ups. In other words, avoid the usual callisthenics course; it will not help your problem. Instead, put yourself on the orthotherapy programme outlined below.

Special Exercises for Low Back Pain

Although some of these exercises duplicate several of the basic ortho-exercises in Chapter IV, they are used in a muscle-strengthening sequence with additional exercises. This sequence should be followed for best results.

This first group of special exercises is to be used during an attack, known as the *acute stage*, of low back pain.

1. Lie on your back with a small pillow under your head, your hands folded on your abdomen, hips and knees bent,

with feet flat on the floor. Keeping one leg in this knees-up, feet-flat position, roll the other knee out and to the side and back up to its original position ten to twenty times. Do this gently and easily, and then change to other leg, ten to twenty times.

2. In the same starting position as in special exercise 1, pull one knee up to your chest with both hands, keeping the other leg in position. Do this, too, gently and easily ten to twenty times. Repeat with other leg ten to twenty times.

3. Again in the same starting position, place your left hand on your left knee, lift your left foot off the floor and move your knee out to the side and round in a circle ten

times clockwise and ten times anti-clockwise. Return to starting position. Repeat with other knee.

Special exercises 1, 2 and 3 represent one cycle. Continue

doing as many cycles as possible, for twenty to thirty minutes at a time for at least two periods a day.

In the *sub-acute stage* of low back pain, continue doing special exercises 1, 2 and 3 and add the following.

4. (A) In the same starting position as before, tighten the seat muscles and tilt the pelvis so that the back flattens

against the floor. Repeat ten to twenty times. (B) This exercise may also be done lying on your stomach with a pillow under the abdomen and a small pillow under the ankles. Just tighten and relax the seat muscles ten to twenty times. Rest. Repeat for five to ten sessions.

5. In the same starting position as exercises 1, 2 and 3, place one hand on each knee and pull both knees close to your chest. If pain is experienced, return to starting position and continue doing exercises 1, 2, 3 and 4. If this knees-to-

chest position is not painful, try rocking the knees back and forth towards the chest easily for one-half to two minutes.

In the *convalescent stage* of low back pain, more active exercises may be attempted. Note that if exercise 5 is still painful, you are not yet ready to go on to this next group.

6. Lie face down on the floor with pillows under the abdomen and ankles, and arms at your sides. Raise your head and one shoulder, and then lower them ten times. Raise and lower head and other shoulder ten times. Do not

raise your legs at all during this exercise. Do this exercise several times a day until it is relatively easy for you to do, then continue doing it at least once a day and add exercise 7.

7. Lie face down with pillows as in exercise 6, but with arms out and in line with your shoulders, elbows bent and fingers pointing over your head. Raise your right arm (keeping it bent) and shoulder blade as far off the floor as you

can, while raising your head and turning it to the right. Then return to starting position. Repeat ten times, then switch to left side and repeat ten times. Go on to exercise 8.

8. (A) In the same starting position as exercise 7, raise one leg, bending it at the knee, and then lower it ten times.

Repeat with the other leg. (B) Now raise and lower one leg with knee straight ten times, and repeat with other leg. When you can do this easily go on to exercise 9.

9. Still in the same starting position, raise your right arm and shoulder, raise your head and turn it to the right and raise your left leg, keeping the knee straight – all at the same

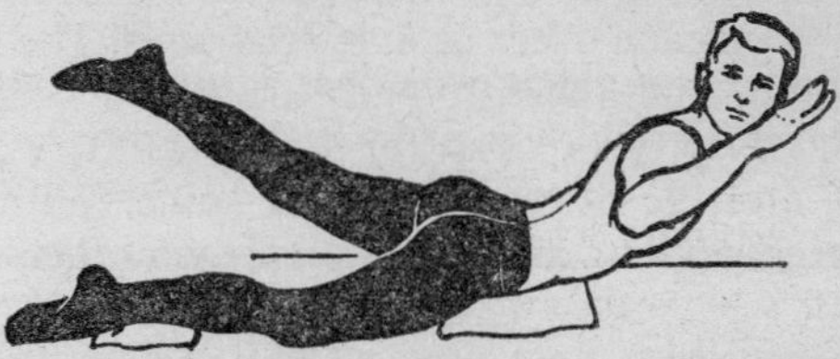

time. Then relax into starting position. Repeat ten times. Repeat by raising and lowering left arm and right leg ten times. Go on to exercise 10.

10. In the same starting position as in exercise 7, 8 and 9, raise your head, arms and upper body as far off the floor as you can, pulling your shoulder blades strongly together and keeping your arms bent. Relax into starting position. Repeat five to ten times.

11. In the same starting position as exercise 10, raise your head, arms (bent at the elbow), upper body and both legs, holding knees straight. Relax into starting position. Repeat five to ten times.

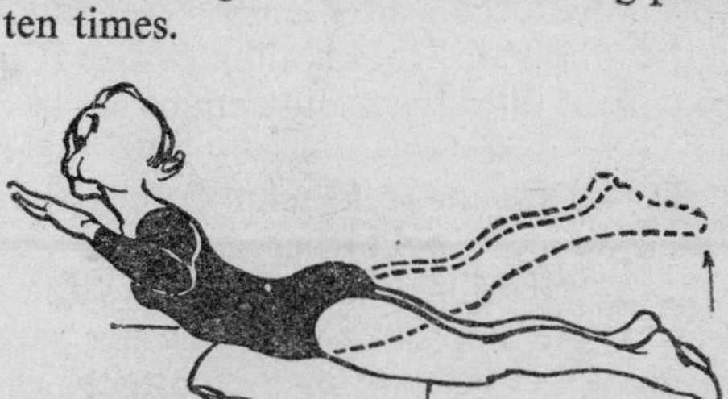

11

10

This is the special exercise programme I give my patients with lower back pain. Once you have completed the special exercises and the pain has abated, go on a maintenance programme of ortho-exercises 7, 8, 9, 10, 11, 12, 14, 15, 17, 18, 19, 20 and 21 as described in Chapter IV. Anyone with chronic lower back pain would definitely benefit from these exercises, but should seek medical advice.

SLIPPED (HERNIATED) DISC AND SCIATICA

Think of the spine as a stack of small cylindrical blocks, separated from each other by small flat pads. The blocks are the vertebrae, and the soft pads are plaques of cartilage called intervertebral discs. The linked construction gives flexibility, and the discs cushion the spine. Without them, the vertebrae would rub and grate on each other with every movement.

Under certain circumstances, a small fragment of disc may slip out of place – a slipped disc. If the disc fragment protrudes from the spine into the soft tissue around it, it is called an extruded herniated disc. Although most cases of slipped disc occur between the ages of thirty and forty, it may also begin much earlier or later.

Sometimes it takes very little to cause a disc to slip. You may simply stoop down to pick up a postage stamp from the floor and, somehow without realizing it, you have moved in such a way that a disc slips out, like a watermelon seed slipping out from between your two fingers. It has even been known to happen when a person sneezes hard. Usually, however, a slipped disc occurs after a heavy weight has been lifted improperly. The pain from a slipped disc can be so severe that the patient falls down and cannot get to his feet again.

Pain from a slipped disc may take many forms. Sometimes the pain is local. You feel it right where it happens, and you may groan 'Oh, my sacroiliac!' At other times the pain is referred. That is, it goes along a nerve to another part of the body, and you may feel it down in the buttocks or lower abdomen, or even in the leg. Pain may radiate right down the leg so that you feel it in your toes, or there may be loss of sensation or weakness in the leg. This referred pain can be quite insidious.

Some individuals 'pop' discs to one side or the other of the vertebral interspace or to the front of it. X-rays show large exostoses or spurs about the vertebral interspace. However, if the disc pops towards the back, it pushes against one or more of the nerve roots of the sciatic nerve. Referred pains result from this pressure being transmitted along the branches of the sciatic nerve.

In a prior chapter we spoke about the manner in which several of the branches of the sciatic nerve course through the psoas major, bringing about muscle imbalance between the psoas major and the hamstrings, and we described how to determine the clinical findings subjectively and objectively.

Sciatica is a neuritis or neuralgia of the sciatic nerve, whose branches run from the hip and back down the back of the thigh to the foot and toes. Popularly, sciatica refers to many different pains that are felt in the hip and leg. These pains are the chain reaction result of iliopsoas muscle imbalance and part of the tibial torsion syndrome. But true sciatica is almost sure to be a result of a slipped disc.

If you suspect that you have a slipped disc, see your doctor at once. If this is his diagnosis, he will probably prescribe something to relieve your greatest pain, as well as an orthopaedic corset extending from the lower rib cage to the pelvis and a belt called a trochanteric strap. Sometimes total rest in bed for ten to fourteen days is necessary, and occasionally pelvic traction will be required. Other possible aids that may be used are the medication to relax muscle spasms, and manipulation or massage to ease the tension of muscles in the area. In severe cases, surgery to remove the herniated disc may be necessary.

Those who have experienced a slipped disc should be particularly careful to avoid lifting heavy weights and should in general follow all the recommendations for low back pain – hard mattress and/or bed boards, knees bent for sitting and sleeping, etc.

Exercises to strengthen the muscles of the lower back area should be a twice-daily routine for all patients recuperating from slipped-disc problems. If you are convalescing following a slipped-disc episode, begin your exercise programme as soon as your doctor permits. At first, follow the same sequence of exercises as those prescribed for the acute stage of low back pain (see page 87). As your condition improves, add the exercises for the sub-acute and convalescent stages of low back pain (see pages 89 to 92). When symptoms abate, go on a maintenance programme of ortho-exercises 7, 8, 9, 10, 11, 12, 14, 15, 17, 18, 19, 20 and 21, placing special emphasis on the two described as follows.

1. Stand on one leg in front of a chair or table that is about the same height as your hips and stretch the other leg straight out, resting your heel on this chair or table, with your toes pointing straight up. Bend your body forward, nice and easy, stretching your arms towards your toes; you will feel a pull in the back of the thigh. Straighten up to starting position and repeat the easy forward stretch five to ten times, then switch legs and repeat five to ten times on the other side. If more pain is experienced after the stretching, wait a few days before trying this exercise again, and concentrate on the low back pain exercises. When you can do this exercise comfortably, repeat it several times daily, especially during attacks of sciatica.

2. Sit on the floor with your legs spread as far apart as is comfortable for you, keeping your knees straight. Bend your body forward, nice and easy, between your legs, stretching your arms out towards your toes. Don't force it. Roll forward, bouncing gently five to ten times. You may want to grip your legs to help pull yourself forward. If you have a belly which seems to get in the way, you can brace your hands on the floor behind you and push from the rear. Repeat this exercise several times daily and aim at getting your torso closer to the floor each session.

ARTHROSIS OF THE SPINE

Another possible cause of pain in the back is arthrosis (misnamed arthritis by many). This is seen mostly in people over forty or fifty years of age.

Physiologically, what happens is that the discs separating the vertebrae start to degenerate. As they age and dry out, their shock-absorber action is lessened, and a stiff and somewhat painful back may result. The spine loses its flexibility and mobility. Some discs may degenerate so much that they are practically nonexistent. When this happens adjacent vertebrae, no longer cushioned from each other, may rub and grate on each other and may eventually even fuse.

Pain may be felt in the groin or in the buttocks, as well as in the back.

Too often treatment of this disorder is based upon X-ray diagnosis. But what appears on an X-ray is the result of Nature's defences over a period of time – defences against abnormal movements of the spine. But the abnormal movements, stresses and strains on the spine are really the result of spino-pelvo-femoral *muscle imbalance.* And this muscle imbalance, ultimately resulting in pain and bone deformity, *could have been detected before the damage was done.*

Long before there is X-ray evidence of arthrosis of the spine, the state of physical unfitness could have been ascertained by dynamic muscle testing of the hip joint and spinal muscles. This is the reason why physical fitness tests should be done routinely to ascertain the early stages of deranged spino-pelvo-femoral muscle balance. Institution of a programme of corrective exercises in the early stages could prevent the arthrosis. Obviously, when the disorder has progressed to the point at which the spinal curvatures or arthrosis are fixed and show up on an X-ray, these effects are not reversible.

In any case of severe back pain, a doctor should be consulted. He is the only one who can pinpoint the cause – slipped disc, arthrosis, arthritis, or whatever else it might be.

For immediate treatment local application of a moist hot pack and deep massage by a physiotherapist are very soothing. If the lesion is confined to a small area, a doctor may decide to inject procaine and corticosteroids. As in the slipped-disc attacks, lying on your side in the knee-chest position and using an orthopaedic corset bring relief. Manipulation to release the tense muscles and myofascial structures is beneficial. Follow the same exercise regime as for lower back pain, if your doctor permits it, including the maintenance programme of ortho-exercises 7, 8, 9, 10, 11, 12, 14, 15, 17, 18, 19, 20 and 21.

SCOLIOSIS

Scoliosis is a sideways curving of the spine.

There is a form discernible at birth, termed congenital scoliosis. The infantile form usually shows up when a baby learns to walk. There is also an adolescent type of scoliosis that is usually noticed between age nine and twelve; it may escape detection because at this age the child bathes himself, and is seldom undressed by his parents.

Scoliosis may also be a delayed consequence of congenital hip dislocation, so the parents of a child born with this disorder should be particularly alert later on for the appearance of spinal curvature.

When for some reason the child with this disorder does not receive early medical care, the problem worsens in adulthood and can result in severely disfigured spinal curvatures and other deformities of the shoulders and hips.

The younger patient may have little or no pain, but by the time he is in his early thirties, the pain can become quite extreme, so that surgery will almost certainly be recommended.

Never attempt to correct a curvature by yourself. This condition must be observed by a physician as soon as its presence is suspected.

Rest in bed and traction are frequently used to treat scoliosis, but often the disorder returns after the person gets out of bed and starts walking. The best treatment is a co-ordinated programme of active and passive stretch exercises.

In formulating the plan of exercises, the orthopaedic surgeon must first determine what muscles are involved.

Usually there are several muscles that are the key to scoliosis, and they must be attacked in sequence:

1. The adductor longus muscle that is contracted and tight must be lengthened and relaxed.

2. Then the semitendinosus is relaxed.

3. Then special exercises to stretch the psoas major and quadratus lumborum are added.

The therapist usually helps by putting the patient through hip stretching and mobilization exercises to a point that just starts to cause pain. Finally, after many weeks, active and passive exercises are used to stretch other related muscles and to bring the ribs back to their normal position.

When the patient conscientiously uses these exercises, he often has no need for surgery, or even for bracing or casting. One thing that may bring relief is a rib belt with foam-rubber lining that is held in position by straps fastening round the back. This belt is worn during the day, counter-acting tensions on the ribs. Corrective body casts or a Milwaukee brace may be indicated.

When should a person have surgery? It depends on the degree of scoliosis and the person's age. If the curve is extreme, surgery will almost certainly be advised. After surgery, braces or casts will help maintain the correction until the spinal joints have adjusted to their new positions. But even when the condition warrants a programme of surgery, ortho-exercises should be used before and after surgery to reduce tensions and stress and to strengthen the various muscles involved.

Do not begin any exercise programme for scoliosis without permission from your doctor. The following ortho-exercises have helped curvature in my patients: 4, 5, 6, 7, 8, 9, 10, 11, 12, 14, 15, 17, 18, 19, 20 and 21.

I have also used the following special exercise programme for my patients with scoliosis. Complete this programme before turning to the basic ortho-exercises.

Special Exercise Programme for Scoliosis

The special scoliosis exercise programme begins with ortho-exercise 10, the long body stretch (see page 61). Do this exercise several times a day until you have achieved a fairly high degree of flexibility (that is, you can press your chest to the floor without too much difficulty). You should also do a variation of this exercise, which consists of assuming the same starting position as the long body stretch and wiggling the pelvis back and forth from side to side. Do this for two to five minutes at a time, several times a day, until you can do it quite easily. Then you can cut down to once a day on the two long body stretch exercises and proceed through the following special exercises, doing each one at least once a day.

1. *Spinal rotator muscle stretch and strengthening exercise*
Kneel on the floor. Put your right knee as far forward as you can and stretch your left leg out behind you. You will be squatting on your right foot. Brace yourself with your left arm, placing your left hand on the floor, and lift your right arm out to the side so that it is on a line with your right shoulder. Bend the right arm at the elbow so it forms a right angle and your fingers are pointing over your head. From this position pull your bent right arm strongly back and turn your head and trunk to the right. The pull on

your right arm and shoulder should be strong enough to turn the head and trunk with it. Then relax into starting position. Repeat pull smartly to the count of four. Rest briefly and repeat to the count of four again. Repeat cycle five to ten times, then switch leg and arm positions and repeat cycle five to ten times with left arm.

2. *Stretching and strengthening exercise for the spinal column and shoulder musculature*

Assume the long body stretch starting position: on your knees, thighs perpendicular to the floor, head and chest resting very near the floor, arms straight out over your head. Keeping your face and chest down and arms stretched out,

shuffle your knees forward one at a time in tiny two- to three-inch 'steps'. Inching forward in this way, move your whole body in a circle on the floor to the right, then in a circle to the left. This may take as much as five minutes per circle.

3. *Strengthening exercise for spinal, shoulder, shoulder blade and hip musculature*

Get down on the floor on your hands and knees. Your knees should be directly under your hips; your hands should be directly under your shoulder joints. Now, simultaneously raise your right arm straight forward, and raise your left leg

straight backward. Try to raise both the right arm and left leg high enough to be on a line with your trunk. Hold this for a count of four, then return to starting position. Then raise left arm and right leg in the same way and hold for a count of four. You will find that your body is inching forward after each lift when you return to starting position. In this fashion you should gradually move across the floor for about three minutes.

4. *Stretching and strengthening exercise for spine, shoulder blade and hip musculature*

Kneel and clasp your hands behind your head so that your elbows fan straight out to the sides. Move your right knee as far forward as you can and stretch your left leg straight out behind you. You will be squatting on your right foot. Your torso will tend to lean forward slightly. From this starting position, turn your head, shoulders and torso

strongly to the right and hold for a count of four. Then move the left knee as far forward as you can, stretch your right leg straight out behind you, turn your head, shoulders and torso strongly to the left and hold for a count of four. You will find that you are again making a series of 'steps' across the floor. Alternating sides, continue moving across the floor in this fashion for from one to three minutes.

5. *Spinal column stretch and abdominal strengthening*

Sit on the floor and spread your legs as far apart as you can. Keep your knees straight. First, bend forward at the waist and stretch your arms straight out over your head

between your legs. Bounce forward in this fashion ten times. Now tuck your right arm behind your back and stretch your left hand towards your right foot, carrying your torso

forward in a stretch. Bounce towards your right foot ten times. Change sides. Tuck your left arm behind your back, reach your right hand for your left foot and bounce ten times.

6. *Stretching exercise for hamstrings, low back and spinal musculature*

Stand with your feet widely separated, legs straight. Bend forward from the waist and reach for the right foot with the left hand, while swinging the right arm straight back and up. Bounce torso in this position, reaching for the right foot with the left hand each time, to the count of four. Return to starting position. Now bend forward from the waist and

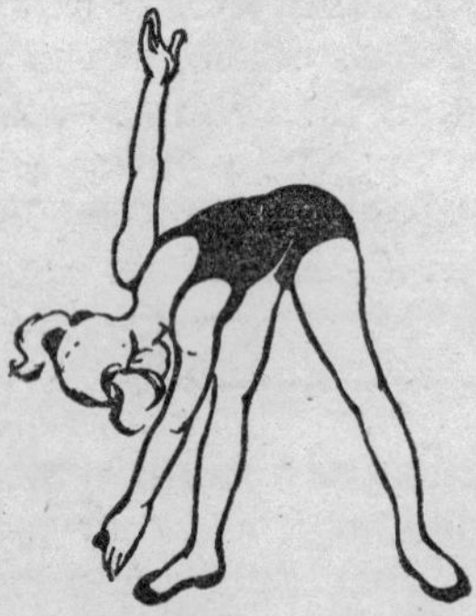

reach for the left foot with the right hand, swinging the left arm behind you. Bounce to the count of four. Repeat this cycle five times.

7. *Strengthening exercise for low back and hip musculature*

Kneel so that your thighs are perpendicular to the floor.

Bend the body forward and rest your face on your folded hands on the floor. From this starting position lift your left leg straight up as high off the floor as you can. Aim to get the lifted leg on a straight line with your trunk. Return to starting position, and then repeat left leg lift ten times. Return to starting position and lift right leg ten times.

8. *Shoulder, back and hip strengthening exercise*

Straddle a low stool. Bend your right leg at the knee and put your right foot on the floor. Turn your torso so you are facing in the same direction as your right leg. Stretch your left leg straight out behind you so that only the pointed toes of the left foot rest on the floor. Hold your left arm close to your side and lift your right arm straight up over your head on a line with your body. Now bend your body backward, bringing your straight right arm with you. Stretch

as far backward as you can and hold for a few seconds. Relax into starting position, and then repeat backward bend five to ten times. Now change sides: left leg forward bent at the knee, right leg straight back; right arm close to your side, left arm over your head. Bend body backward again five to ten times.

9. *Abdominal and trunk rotator strengthening exercise*

Lie on your back on the floor with your arms stretched straight over your head. Tighten your seat muscles so that the small of your back is flattened against the floor. Holding this position, raise your left arm and shoulder and your head and turn your upper body strongly to the right for a count of four. You are aiming for a strong enough twist in the trunk to roll your upper body over on to your right shoulder while keeping your hips and legs flat on the floor. When this exercise is done correctly the lower ribs will be pulled in and the abdomen will flatten considerably on the

side of the arm being lifted. Repeat the twist to the right five or ten times. Return to starting position. Now lift right arm and shoulder and head, turn strongly to the left and hold for a count of four. Repeat five to ten times.

There are numerous other exercises – both stretching and strengthening – which may be used in correcting scoliosis, but these are the nine exercises I use most often. They are difficult, but well worth the effort.

SYNOVITIS OR BURSITIS OF THE HIP

These are common disorders of the hip. *Synovitis* is an inflammation of the membranes lining a joint. *Bursitis* is an inflammation of a bursa (a bursa is a fluid-filled cavity that surrounds and protects the adjacent muscles or tendons). Synovitis is a disorder that usually occurs in children, while bursitis most frequently occurs in adults. However, the symptoms of this pair of ailments and their treatment are similar.

The patient usually limps, may have pain, and there is usually tenderness in the hip joint in response to pressure. In an acute attack the pain can become intense, and a muscle spasm (of the iliopsoas) will force the leg into a flexed position.

Clinically, there is a minimal temperature rise. X-rays may show capsular thickening of the hip joint. This ailment may be difficult to detect, especially during its non-acute stages, but performance of the dynamic tests as outlined in Chapter III will provide a conclusive diagnosis. Only thorough examination by a physician can determine the specific cause of the attack and management of the treatment.

The usual treatment for bursitis or synovitis starts with twelve to fourteen days of rest in bed. Exercises should be done only at the recommendation of the treating physician. Usually, after the acute phase has passed, exercises of the type given in the acute stage of low back pain are recommended (see page 87). When symptoms abate, I have found a maintenance programme of ortho-exercises 7, 8, 9, 14, 15, 17, 18, 19, 20 and 21 very useful.

CHRONIC HIP PAIN

Pain in the hip and lower back can also be due to one leg's being even an imperceptible bit shorter than the other. Many sufferers from severe back and hip pain have found sudden, seemingly miraculous, relief when a half-inch heel pad elevation was placed inside or on the heel of the shoe worn on the foot of the short leg.

Other treatment for hip pain includes hot packs and deep massage, local injection of hydrocortisone or procaine, or muscle relaxant medication may be prescribed by your doctor.

Exercises to increase flexibility of hip movement are the same as those recommended for lower back pain. A maintenance programme of ortho-exercises 7, 8, 9, 14, 15, 17, 18,

19, 20 and 21 is also strongly recommended, once symptoms abate.

ARTHROSIS OF THE HIP

Degenerative disease of the hip joint that usually appears in patients past the age of forty is a form of osteo-arthrosis or osteo-arthritis.

Osteo-arthrosis may affect the arms, knees, shoulders, spine and fingers as well as the hip joint – that is, any joint may be affected. The onset may be due to ageing, injury or progressive muscle imbalance. Adult arthrosis of the hip can also be a result of childhood hip disorders that were neglected or treated inadequately.

In this ailment the cartilage in the joint softens and degenerates, resulting in the wearing away of the cartilage. The bone creates new tissue which is then laid down, making the joints swollen and knobby.

Long before manifestations of pain or infirmity, the patient whose osteo-arthrosis is affecting his hip joint will walk with one foot turned outward – an early attempt to compensate for stiffness in the joint. He may not even be aware of this unless it is brought to his attention. Dynamic testing for the signs of hip tension, as described in Chapter III, will also disclose the presence of osteo-arthrosis at the earliest possible time. This is the golden opportunity for an intensive ortho-exercise programme with an excellent possibility of reversing the disorder.

In the early stages, pain appears only during walking or strenuous exercise. Later, the pain may become much more intense, and it may radiate down from the hip to the groin, thigh and knee. The patient begins to walk stiffly and awkwardly to compensate for the pain; external rotation or turning out of the foot is commonly found. As the bone deterioration progresses he may find that he can no longer extend his leg with the knee straight, rotate his leg to turn his toes in or cross his leg to put on a shoe, trousers or stockings.

X-rays may show nothing abnormal in the early stages,

but diagnosis, although difficult, may be ascertained by the hip tension signs. Nevertheless, if arthrosis of the hip is even suspected, consult your physician regarding corrective exercises. The telltale signs of bone degeneration will show up on X-rays later, but the ortho-exercises may be effective before that stage of the disease is reached.

Treatment must be individually designed for each patient. Resting the hip too much encourages stiffening, while over-activity brings on strain.

Ortho-exercises to increase flexibility of hip movement are the same as those recommended for low back pain (see pages 87–92). A maintenance programme of ortho-exercises 8, 9, 13, 14, 15, 17, 18, 19, 20 and 21 is also recommended once symptoms abate. First, exercises to increase the range of motion of the hips (low back pain exercises) should be performed ten to fifteen minutes at a time, two or three times daily.

Other aids that may be helpful in individual cases are aspirin, heat, massage, manipulation, a cane, a heel lift and a firm mattress. The patient should lie flat on his back for thirty minutes in the middle of each day, for complete rest. But these measures relieve symptoms only. Exercising is the only way to stretch the muscles and alleviate the tension putting stress on the joint.

In some cases, medication may be required to relax the muscles enough to permit exercise to be effective. Your doctor may recommend the use of a special strap to stabilize the hip joint during active exercises.

Never attempt to treat a suspected case of arthrosis of any joint yourself. Only a doctor's diagnosis can pinpoint the cause of pain in any joint, and all treatment should be under his guidance.

TORN LIGAMENTS OF THE KNEE

We will not concern ourselves here with direct trauma or injuries that tear the ligaments of the knee. In the vast majority of ruptures of the ligaments connecting the leg muscles to the knee joint, a prior chain of events has made the knee vulnerable. A pre-existing condition of musculoskeletal weakness and imbalance must have set the stage and created the vulnerability, or the knee would not have given in to injury.

In my experience, this injury is most likely to occur in those individuals who exhibit the tibial torsion syndrome described earlier. To compensate for the strains of their imbalanced musculature, these thirty per-cent-ers have developed such secondary symptoms as flat feet, high instep, hyper-extended knees, knock knees and bow-legs. Walking and more strenuous movements, for people with these symptoms, results in torque stress at the level of the knee or in the foot or both, depending on the position of the knee during gait. Wearing improper, ill-fitting or inadequate footwear, or cleated shoes during active sports or even walking, may set the stage for a pivotal strain and laceration of the ligaments of the knee.

When a ligament ruptures, there is tenderness over the entire area. The knee joint becomes bruised, swollen and weak, and there may be a sensation of 'catching' within the joint when the knee is moved.

If you have had such an injury, you may be impressed by the physician's moving your knee in a 'wobbly' direction. But what is more important to note is that when you stand naturally your knee and foot do not point in the same direction. If you try to hold your leg so that your knee is facing front your foot will be turning out. If you point your foot straight ahead, your knee turns in. The 'twist' from knee to foot creates a strain. Any sudden bending of the knee, especially when the knee is supporting the body's weight, as in squatting, may prove too much for the already strained knee ligaments.

Immediate treatment should include ice packs around the knee and elevations of the leg.

When ligaments are torn, casting or surgery are necessary for healing and return to normal functioning.

When the patient is able to walk again, a knee brace should be used to prevent sideways movement or hyperextension; the heel of the injured side should be raised half-an-inch to prevent strain and stress on the injured ligament.

Ortho-exercises should be started as early in the convalescent stage as possible. They will strengthen the knee and leg muscles and correct the musculo-skeletal imbalance that follows such injuries. Only start exercising under a physician's supervision. I have often used the special knee exercises that are described below. After symptoms abate I recommend a maintenance programme of basic ortho-exercises 13, 14, 15, 16, 18, 19, 20 and 21.

Special Knee Exercises

1. *Knee-cap shrugging*

Sit on the floor with your legs pointing straight ahead and your knee caps pointing straight up towards the ceiling (this does not mean that your feet will also point straight up, as we have just explained). Brace yourself by putting both hands on the floor behind you. (A) In this starting position

quickly tighten and release the top muscles of the thigh as many times as you can. This may be done thousands of times a day in order to build up the responsiveness of these muscles. You should see your knee caps 'jump' a little each time you tighten them. (B) In the same starting position, tighten the top thigh muscles, and hold the contraction for a count of from ten to thirty. This is more tiring, as it brings more muscle fibres into action, but it is very beneficial.

2. *Vastus medialis strengthening exercise*

(The vastus medialis is the muscle that connects the knee cap to the inside of the thigh; it forms the prominence inside the thigh just above the knee joint.) Sit on the floor with your legs stretched out in front of you, knees facing straight up, as in knee exercise 1. Place a padded block four to six inches high under the right knee. Place a weight over the right ankle. The weight can vary from one to fifty pounds, depending on the strength of the muscle. Start with one pound, perhaps a sock filled with sand, and work your way up. With the one-pound weight over your ankle, lift your right foot and straighten out your right leg.

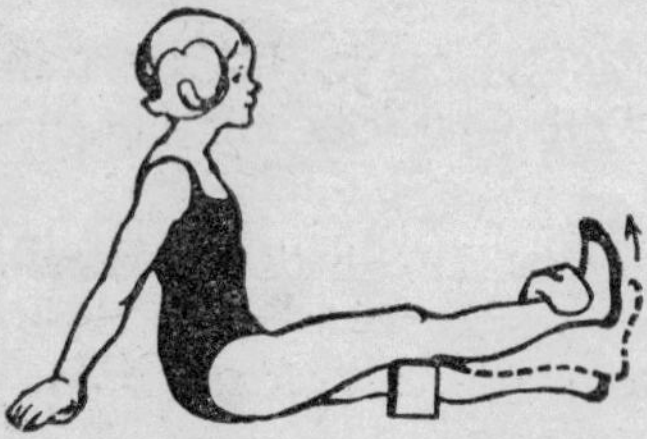

Relax into starting position. Lift right leg ten times. Rest, then add a few more pounds of weight to the right ankle and lift another ten times. Now rest the right leg while you shift the block to under your left knee and tie the one-pound weight to your left ankle. Repeat the same cycle with the left leg. You can continue repeating this cycle every day, adding a few more pounds each time.

3. *Hamstring power-building exercise*

Lie face down on the floor with small pillows under your ankles and abdomen and a thin pad under your knees for protection. You may want to rest your head on your folded arms for comfort. (A) Tie a weight of one pound to your right heel. Keeping your right thigh on the floor, bend your right leg at the knee and slowly raise your leg so that it is perpendicular to the floor. Then lower it to starting position. Lift in this fashion ten times. Rest, and then, increasing the

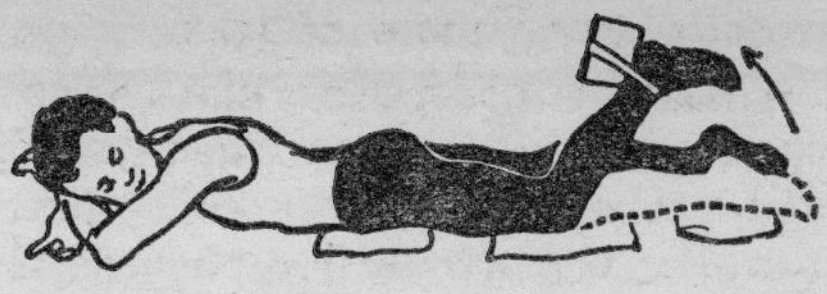

weight on the right ankle by a few pounds, raise right leg ten times again. Rest right leg and switch one-pound weight to the left leg. Repeat the whole cycle with the left leg. As in exercise 2, repeat several cycles of this exercise every day, gradually adding weight up to fifteen pounds. (B) From the

same starting position and with the same progression of weight attached to each ankle, raise each leg, holding it straight, as high as you can off the floor, at least ten to twenty times once a day.

CALCIFICATION OF THE MEDIAL LIGAMENT OF THE KNEE

A fairly common condition in men between the ages of twenty-four and forty years, is hardening or calcification of the ligament at the inner side of the knee joint (Pelligrini Steida disorder). Men who ride a lot and who are thirty per-cent-ers are vulnerable to the torque or twisted pressure of the inner knee against the saddle. Bony growth develops at the points at which the ligaments attach to the knee joint.

The patient can bend his knee, but when he tries to extend it completely, he experiences pain. There is also tenderness when pressure is applied to the middle of the knee.

Check with your doctor about exercises to increase the knee's mobility and strength and about appropriate footwear. The same special knee exercises for torn ligaments of the knee are recommended. Occasionally, the bony part of the ligament needs to be removed by surgery, but this is almost never necessary if the ortho-exercise programme is faithfully carried out. Virtually the same maintenance programme as that recommended for torn ligaments of the knee is recommended to correct the musco-skeletal imbalance usually at the root of this disorder, that is, ortho-exercises 7, 8, 13, 14, 15, 17, 18, 19, 20 and 21.

SWELLING OF THE KNEE

Any repeated swelling of the knee should be reported to your physician; only he can diagnose the cause of this symptom. If the swollen knees regularly occur after abrupt change from high-heeled shoes to bare feet or slippers, or after walking on rough terrain, ortho-exercises are helpful for long-term relief.

The person who has proved vulnerable to knee injuries in the past should avoid wearing flat shoes, plimsolls, slippers, sandals or cleated shoes. He should not go barefoot. He should always wear supportive shoes with extended counters and adequate heel pads.

Special exercises recommended for all knee joint weaknesses are those described above for torn ligaments of the knee, including a maintenance programme on ortho-exercises 7, 8, 13, 14, 15, 17, 18, 19, 20 and 21.

RUPTURED MUSCLES AND TENDONS OF THE LEG

These ruptures occur in athletes – tennis players, skiers, sprinters, hurdlers, and baseball, football and basketball players.

The rupture may be traced to an injury or, more frequently, may occur spontaneously in the thirty per-cent-ers wearing sneakers or other forms of inadequate compensating footwear. The calf muscle may rupture if it contracts suddenly when the foot is held squarely on the floor. This is felt as a sudden sharp pain in the calf with a loss of support and a sensation of something having snapped. Pain may be accompanied by bruising and swelling; sometimes one can actually feel a gap between the muscle fibres.

This kind of accident is seen especially in individuals of stocky build, who have been wearing flat shoes with no support.

Treatment by your physician is essential. A mild or partial rupture of the muscle or tendon of the leg may be held in place by a cast for several weeks until it heals. Supportive shoes with a heel pad three-eighths of an inch high should be worn. In a severe tear, muscle or tendon fibres must be sutured together to ensure return of muscle strength.

Exercises to strengthen the recuperating muscle as well as the other leg muscles should be done as directed by your physician.

Special Leg Exercises

I recommend the following special exercises for my patients with ruptured muscles or ligaments of the leg:

1. *Static contractions*

These are muscle-setting exercises in which no motion takes place in the joints. Only a tightening of the specific muscle involved occurs. This may be done, while sitting in a chair, from 100 to 200 at each sitting, several times daily or even more.

2. *Non-weight-bearing exercises*

(A) Sit on a table that is high enough to allow your legs to dangle freely. Kick the affected leg forward and back fifty to 100 times. (B) Sit on a table with your legs dangling or on a chair with both feet flat on the floor. Keeping your heels on the floor, raise your toes and forefeet as far off the floor or up in the air as you can, then relax into starting position. Repeat ten to twenty times. (C) Sit on a table with your legs dangling freely. Twisting your ankles, make circles with both feet in an inward direction ten to twenty times and in an outward direction ten to twenty times. All of the last three exercises can be done with toes straight or bent. (D) Sit on a chair with your feet flat on the floor. Raise both heels off the floor as high as you can, keeping your toes on the floor, then lower them until both feet are flat on the floor again. Repeat twenty to thirty times.

3. *Weight-bearing exercises*

(A) Stand with your back to the wall, leaning against it, and your feet approximately one foot away from the wall. Keeping your heels on the floor, raise the forefeet and toes off the floor and then lower them ten times. (B) In the same starting position as A, raise and lower the inner borders of both feet ten times. (C) In the same starting position again, but with the feet held about sixteen inches apart, raise and lower the outer border of both feet ten to twenty times. (D) Face the wall, standing erect, rise up on your toes and then come down ten to twenty times. (E) Face the wall and stand erect on the affected foot only. Now lift the heel of this foot off the floor, rising on your toes. If you find this impossible, start up on the toes of both feet. Now lift the unaffected leg off the floor while lowering the heel of the affected foot. This is a way of practising to stand on the toes of one foot. Do this exercise so that you are standing on the toes of the affected leg (either right or left) and lifting the unaffected leg. Support the hands on a table or chair for balance if needed.

Once symptoms abate, my patients go on to a maintenance programme of ortho-exercises to correct the underlying musculo-skeletal imbalance which is at the root of their discomfort. I recommend ortho-exercises 13, 15, 16, 19 and 20.

SPRAINED ANKLES

Although sprained ankles can occur following an accident, our concern here is the 'chain reaction' sprained ankles of the thirty per-cent-ers. The vulnerable thirty per-cent-er may constantly resprain his ankles until his ankles become weakened and 'give way'.

When put under sudden stress, the ligaments – usually of the outer side of the ankle – may lacerate or the tendons of the ankle can easily slip out of the grooves in the bones that hold them. Typically, a skier swoops down the slope, slips and turns his ankle with real force, lacerating the outer ligaments of the ankle or jarring the tendon out of place.

The injured ankle should be taped or strapped for about six weeks. A cast may be needed until the lacerated ligaments mend. The out-of-place tendon may pop back into place by itself, or strapping or a cast may be used. Sometimes surgery is necessary to suture the tendons firmly into place or to deepen the grooves where the tendons lie to give them more stability.

Corrective footgear, as for any tibial torsion syndrome disorder (see page 149), and reasonable caution should prevent recurrences.

When your physician finds you ready for exercises, non-weight-bearing and weight-bearing exercises may be taken as prescribed under 'ruptured muscles and tendons' (see pages 113 and 114).

BUNIONS

Strange as it may seem, bunions are also caused by muscle imbalance. I believe that they are adaptations to the tibial torsion syndrome and usually occur in the thirty per-cent-ers. When the legs and feet are forced into poor standing and walking positions, various parts of the toes are pressed against the sides and top of the shoe. Bunions form at these pressure points.

Bunions are common only where people wear shoes. Women's feet are more frequently afflicted than men's because the highly styled shapes of their shoes, with their inclined arches, are more constricting.

Bunions can be removed by surgery, or can be relieved by various local salves or lotions. A cushion to cover the sore area will prevent it from rubbing against the inside of the shoe. A looser shoe will help, too. But these are only temporary measures. Unless exercise therapy and appropriate footwear are provided to correct the muscle imbalance that is the root of the problem, bunions will inevitably recur.

CALCANEAL BURSITIS

Sometimes pain felt in the heel of one or both feet is due to the inflammation of bursitis, or to bony spurs that grow out of the heel bone and project down into the heel, or to ingrown corns. The pain is especially likely to show up in the thirty per-cent-er, either in adolescence or between thirty and thirty-five years of age. The pain occurs during standing and increases in intensity after prolonged walking or standing on hard surfaces for a long time.

Placing a sponge rubber pad half-an-inch thick in the heel of the shoe helps to raise the heel and acts as a cushion. Consult your physician about corrective footwear.

SORE FEET

When your feet hurt and there is, to your knowledge, no specific cause you can ascribe it to, it's a safe guess that the related weakening of muscles in the hip, leg, ankle and foot is the problem. We have explained how abnormal outward rotation of the hip (often caused by a tightened or shortened iliopsoas) can place undue stresses and strains on the muscles of the legs and on the knee and ankle joints. At the end of the line are the muscles and ligaments of the feet.

Sometimes the foot pain is felt in the arch; sometimes the entire foot aches. At the age of ten or fifteen, pain may be felt especially in the ball of the foot after the child has been on his feet a lot. Women frequently get cramps in the arches after switching from high-heeled shoes to flats or bare feet.

Quite often years of suffering from sore feet will be found to be due to one leg being slightly shorter than the other. In these cases a lift in the heel of the shoe of the shorter leg will cure the problem overnight. But in cases where the muscles are at fault, exercises to bring the leg and foot muscles into proper balance will take away the strain and the pain.

When sore feet are due to weak or tight muscles of the hip, knee, ankle, feet and/or toes, exercises to stretch and/or strengthen the muscle components of the above regions may be instituted. Special exercises for the hip, knee and leg have been previously described. I often recommend ortho-exercises 13, 15, 16, 19 and 20 to my patients with sore feet. (Exercises 19 and 20 concentrate on loosening the iliopsoas, which is usually the real culprit behind chronic sore feet.) The following are some special exercises for correcting foot problems.

Special Foot Exercises

1. *Non-weight-bearing*

(A) Sitting on a chair with your feet flat on the floor, curl and straighten your toes ten to twenty times. (B) In same starting position, curl your toes and lift the inner borders of both feet at the same time. Relax feet into starting position. Repeat ten to twenty times. (C) Sit on a chair with your feet flat on the floor. Place two boxes, one full of marbles, the other empty, right in front of your feet. Pick up the marbles with the toes of one foot and transfer them to the empty box. Alternate feet. (D) Sit on a chair with your feet flat on one end of a turkish towel spread on the floor. Pull the towel under your feet by curling your toes. Straighten the toes and curl them, pulling the towel farther under the arch of your foot each time you curl your toes. Continue until the whole towel is curled into a roll under the arches of both feet. (E) In the same starting position as D, simultaneously curl the toes of one foot and raise its inner border, while turning the foot inward, then follow suit with the other foot. This time you will find that you are both curling the towel and pulling it into a sort of ball between your feet. Continue until the whole towel is curled up between your feet.

FLAT FEET

Flatfootedness can be due to heredity, injury, obesity, muscle imbalance or extra little bones in the foot, or to a combination of all of these. In flatfootedness, the arch is flat instead of curved upward; it may be so low that the entire sole of the foot rests on the ground. Sometimes this condition, which initially affects only the soft tissue, gradually involves the foot bones as well.

Many types of flatfootedness can be recognized in early infancy, but the flatfootedness of the thirty per-cent-er usually shows up when he starts to walk. Other forms do not become apparent until he is in his teens. With this type of flat foot, there appears to be a normal arch when the foot is raised, but when the person stands, the arch squashes down as flat as a pancake.

Exercises and corrective shoes with moulded supports will relieve pain. Exercises to strengthen the muscles of the foot and leg are the same as those recommended for sore feet.

FATIGUE FRACTURES OF THE FOOT

Stress or fatigue fractures of the small bones in the foot can occur when there is so much tension on certain bones from muscle pull that the slightest additional force will break or dislocate the bone.

The excess muscle stress is often the result of tibial torsion syndrome.

Such fractures are especially common in military recruits during basic training and can usually even be pinpointed to the week of training when long marches begin.

Sometimes symptoms may be absent at the time of the fracture but will show up a week or so later. A similar condition is familiar to ice skaters and is, in fact, known as 'skater's fracture'.

After the fracture has healed, only support shoes should be worn. The muscle-strengthening exercises recommended for my patients with tibial torsion syndrome may also be used: that is, ortho-exercises 12, 14, 15, 19 and 20. Of course, consult with your physician before beginning an exercise programme after recovery from a fracture.

MUSCLE FUNCTION AND SEX

If your body is racked by aches and pains all day, you're simply not going to be in the mood for sex at night. Even if you aren't in severe pain, your body may be under a severe strain that leaves you completely worn out at night, just too tired to do anything, including making love.

Some people are very much aware of what their problem is. When the back or hips hurt, even during normal walking and sitting, they won't be less painful during the specific and vigorous movements of love-making. But sometimes good hip movement may be hampered without there being any warning symptoms of pain. The legs just won't stretch out.

A person with well-balanced thigh muscles can sit on the floor with his legs 130 to 140 degrees apart, or he can lie on his back with his knees bent and his feet in the air. The legs should be supple enough to spread the knees apart widely. A woman who does not have this flexibility of movement will find intercourse painful and develop other sexual problems.

Sometimes women experience back or hip-area pain that is directly traceable to a sexual experience. I remember during World War II an officer's wife who had been in several military hospitals for pain in her lower back. She had first experienced the pain immediately after a vigorous sexual session, and it persisted off and on for several years. Sometimes she would feel it in the lower part of her back and sometimes it radiated into the hip area. It was always most severe following intercourse, and naturally after this time her marriage was being affected. She had been examined by gynaecologists, internists, surgeons, neurologists, urologists, and psychiatrists. None had discovered any pathological condition that could have caused her discomfort.

We suggested that she place a very firm pillow under her buttocks during intercourse to keep her pelvis tilted at an angle to reduce muscle strain. We recommended a special exercise programme, which we have used as well for other patients with related problems. Within a month, she reported that her pain had largely disappeared, and we were able to tell her that after another month she could cut the exercise schedule back to a maintenance dosage of fifteen minutes every evening. Of course, if the pain recurred, she was to resume the full exercise schedule.

The four special exercises to condition muscles needed for a woman's active participation in sexual intercourse are:

1. *Adapted fencer's stretch*

Get into a fencer's thrust position. Place your right foot forward, bending your knee and stretching your leg as far in front of you as you can. Turn your right foot in slightly. Stretch your left leg out straight behind you so that your foot is braced against the floor with the heel off the floor. *Turn the foot of this leg strongly in.* Hold your torso erect and bend back from the waist slightly. Bounce your torso backward until a pull is felt in the groin. (You may want to put your left hand on your left hip and rest your right hand on your right thigh to help balance yourself.) Do fifty bounces on your leg, then switch positions so your right leg is back and do fifty bounces. Repeat several times a day.

2. *Adapted hamstring stretch*

Stand on one leg in front of a table or bench that is about as high as your hips. Turn the foot of this leg strongly in.

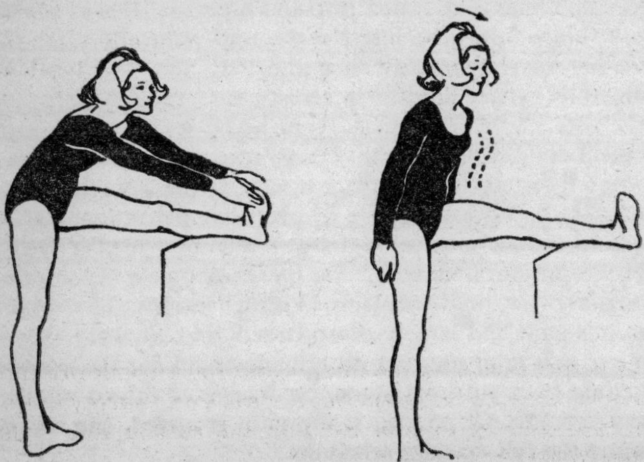

Stretch the other leg straight out, resting your heel on the table with your toes pointing in. (A) Bend forward from the hip, reaching towards your outstretched foot with both hands and bounce your torso hard, trying to reach farther towards your foot each time. Bounce 100 to 200 times, then switch to the other leg and bounce 100 to 200 times. (B) If you are round-shouldered, and/or have a rounded upper back, assume the same starting position, but bounce the torso towards the outstretched leg while holding your spine straight, shoulders back, and head up in a 'good posture' position, arms at your sides. Bounce 100 to 200 times towards each leg.

3. *Hamstring stretch for outer range of motion*

Lie on your right side on the floor in a fairly relaxed position, bracing yourself with your right hand. Grab your left heel with your left hand and stretch your left leg straight up into the air, so that the knee is completely straight and the

leg is perpendicular to the floor. Hold for a count of four, then relax into starting position. Repeat ten to twenty times, then turn over on to your left side and stretch your right leg ten to twenty times.

4. *Adductor stretch* (see page 67)

The basic ortho-exercises that help increase the hips' range of motion are also useful. They are ortho-exercises 7, 8, 9, 14, 15, 18, 19, 20 and 21.

MUSCLE IMBALANCE AND PREGNANCY AND MENSTRUATION

Many complaints associated with a woman's menstrual cycle are due to muscle imbalance. Many women get swelling of the knee and pain in the knee joint during their menstrual periods. Backache at this time is common, too. I have patients whose attacks of headache, low back pain and often radiation of pain into the lower limbs correspond to the onset of menstruation every month.

In many cases, doctors are unsure of why these symptoms show up only at this time. However, it has been suggested that such menstrual problems might be caused by a circulatory disturbance in the pelvic area, due to congestion of the major blood vessels, which in turn causes tension in the muscles not only in the pelvic area, but also radiating into the leg or up into the shoulder. But I have noted a correlation between poor posture or muscle imbalance and severe menstrual discomfort. Among the body's adaptations to muscle imbalance in the kinetic chain reaction is a forward tilting of the pelvis. This results in an increased horizontality of the sacrum, which in turn places unusual stresses on the low back area, and also in the shoulder region. When there is considerable abnormal tilting of the pelvis, the abdominal organs sag because they are not being supported properly by the musculo-skeletal system. This adds to the pressure and pelvic congestion which occurs during the menstrual period.

Correcting the postural defect by exercising the appropriate muscles eliminates the pressure and pelvic congestion caused by sagging abdominal organs, and can help bring relief of the menstrual symptoms as well.

This theory may also account for some of the similar symptoms that often show up in pregnancy. A mother-to-be may have pain in her back, thighs or knees. The lower back pain of pregnancy occurs because, when the pelvis is supporting its unaccustomed burden, the muscles are under unaccustomed strain. This condition is usually worsened by wearing high heels, which exaggerates the tilt of the pelvis. Pregnant women should wear safe and comfortable walking shoes, which will help prevent falls as well.

Some muscle upsets are normal during pregnancy. If you are innocently standing at the kitchen sink scraping carrots and your hip kicks out and gives way so that nothing is holding you up but that carrot waving in the air, don't worry. The kicking out occurs because of changes in hormones near the end of pregnancy that make a major ligament relax and 'let go'. When this happens to you, try not to fall, but don't be too concerned. Your muscles will 'get it together' in a minute or two.

Postural defects can be a significant cause of the more painful symptoms of menstruation and pregnancy. Low back pain at either of these times is neither normal nor necessary. And we often have proof of this when, after a programme of corrective exercises, a young woman reports that her menstruation-related pains have disappeared along with the symptoms of sore feet that she came into the office to correct in the first place. Why does this happen? We theorize that, when we strengthen the hip muscles for their primary job of supporting the body in proper balance for walking and standing, they are strengthened enough at the same time to support congested blood vessels or a distended uterus. In any case, the proper tone of the 'psoas shelf' triggers a correction of the kinetic chain of muscles right down to the calf muscle's attachment to the heel bone.

Consult with your physician, and if he agrees to a programme for the relief of menstruation-related symptoms, ortho-exercises 7, 10, 14, 17 and 19 are the ones I find especially helpful. Pregnant women should go on a modified exercise schedule, fifteen minutes a day, of ortho-exercise 11, and any selection of exercises the doctor recommends. (Special scoliosis exercises 1 and 2 will also help these conditions.)

CHAPTER SIX

Testing Muscle Imbalance in Infants and Children

The best time to uncover a problem of muscle imbalance is when a child is very young. At this early stage before secondary effects have a chance to occur, fairly simple measures can often cure the problem in a short time, and will certainly prevent permanent damage.

A careful, thorough physical examination should be made on every baby at birth and again in the first few days of life, usually while the infant is still in the hospital. Every infant should be seen by a paediatrician every month for the first few months of his life, every three or four months until the child is two or three years old, and every six months until he is eight or nine. He should have yearly examinations after that. At each one of these visits, the paediatrician should check the development of the child's muscle and skeletal system. These few minutes for an orthopaedic survey can make all the difference in the world to the child's healthy future.

Beyond seeing that your child gets proper medical attention, you yourself should also be alert to indications of a problem. If a doctor must be a Sherlock Holmes to find muscle and skeletal disorders, you must be the Doctor Watson, keenly keeping a lookout for clues that you can report to the doctor. Very often it is an alert parent who calls the doctor's attention to a situation he might not otherwise notice.

The physical tests to determine muscle problems in infants and children are simple and are very similar to those outlined for adults described in Chapter III.

These tests will not let you diagnose specifically what is wrong. That is the doctor's job. But the tests are designed as a screening device to give you a warning signal, to tell you that something is wrong somewhere. You should always see a doctor for advice on a specific problem and what should be done about it.

The tests should be repeated periodically (not every week or every month, of course; but on about the same schedule as the child's regular physical examinations). This is because, although your child may appear to be perfectly normal at an early examination, some of the signs that were not far enough developed to show up earlier may become noticeable as he grows older. The growth of the muscles sometimes does not keep up with the growth of the bones.

Some Signs Can Be Detected Without Tests. In his first few weeks of life, an infant's joints are normally flexed or slightly bent, and move rather stiffly. So don't worry when you see that the arms and legs of your newborn baby don't extend completely. These joints will loosen up gradually in the next few weeks. Then, however, if one joint seems to remain stiff for a considerable time after all the other joints seem to have loosened up, or if one arm or leg is still bent while the other extremities can be completely extended, then you should call this to your doctor's attention.

If there is any swelling, or pain, or redness, or heat at any one spot, you should report it to your physician. Look for any tightness of the skin, especially in the creases. All skin folds should be examined to see that the skin is supple and moves freely over the underlying muscle or bone.

In some cases an observant parent notices signs of muscle and skeletal disorders in normal day-to-day caring for an infant, even without doing any tests. One mother who came to us explained that she had difficulty putting a nappy on her child because one leg and sometimes both legs would rotate outward. Another mother said maybe she was being silly, but one thigh of her baby simply 'looked different from the other'. She wasn't silly at all. The child had a dislocated hip.

Another mother noticed that her child did not sleep in the typical foetal position with its legs bent, but had a sleeping posture that somehow seemed different.

Preliminary Examination of an Infant. Before doing the tests, place the infant on a table or counter with a flat, hard surface. You can put a soft towel or blanket down on the table to make it a little warmer and cosier. Always do the tests very gently, and never force the joints of a baby or older child. And, above all, never leave the child on the table or counter alone for an instant. It is a good idea to have someone read directions to you and hold the illustrations up for you while you perform the test. Then you do not have to turn away from the child even for a moment. Even a very young baby can squirm round in a split second and fall, with disastrous consequences.

With the baby lying on his back on the table, compare the lengths and circumferences of the two arms and the two legs to see if they appear to be equal. If one is smaller, it quite likely indicates lack of development of the muscles or a related problem. Turn him over and examine his spine carefully for any signs of curvature or crookedness. Also, look for tufts of hair, dimples in the skin or discoloration along the spine. Feel the spine with your fingers for apparent blank spaces in it. Any of these findings should arouse suspicion of some type of abnormality in the spine area.

With the baby lying on his back, look at the folds and creases on his thighs and see if the creases are equal and even. Are some deeper and more prominent than others? Hold his feet up as though you were putting on his nappy and look at the creases again. Turn him over on his stomach and look at the folds of his buttocks and thighs. They should be even and equal at the back, too.

Now place your hands under his armpits and hold him up in the air. Do his legs and feet hang normally and equally, or does one leg seem to be shorter or more bent than the other?

Another way to compare the lengths of the two legs is as follows: with the infant lying on his back, bend his legs at the knees and place his feet flat on the counter or table near his buttocks. The knees should be level with each other.

If one is higher than the other, it might indicate that one leg is shorter than the other or that the hip is dislocated.

Tension Tests. With the baby lying on his back and his legs extended, gently press the knee joints downward against

the table. Do legs or feet turn inward as shown in the illustration? This indicates tibial torsion syndrome.

The test for tension of the iliopsoas muscle in infants is similar to the one used to test adults. The baby lies on his back on a table, with his buttocks close to the edge of the table. Hold one of his legs, gently bend it and place it across his abdomen. This will hold the pelvis firmly and gently in place while keeping the spine against the table. Now with your other hand grasp the infant's other knee and, always gently, push that leg down past the edge of the table. If hip

and back muscles have normal length and flexibility, this leg can be fully extended. If the muscles are short or rigid, or otherwise abnormal, the free leg will not extend. The free leg may even rise off the table and into the air when the other foot is placed against the abdomen in the test position. Repeat for other leg.

This next test starts with the infant on his back with knees bent and feet placed close to the body. With one hand on each of his knees, gently spread the legs apart, keeping the knees bent and the feet close to the buttocks. A normal

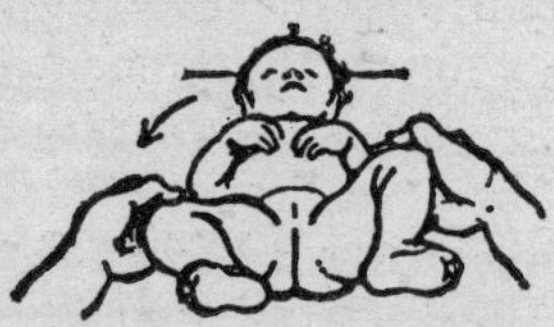

leg can be pushed to the side until the thigh lies flat on the table. If there is a muscle abnormality, the leg will not extend this far, but will show resistance about halfway down as shown in the illustration.

If the child is old enough to stand, put him on his feet and notice his posture. If his feet and legs are spread wide apart and his feet turn outward, if his knees are not straight

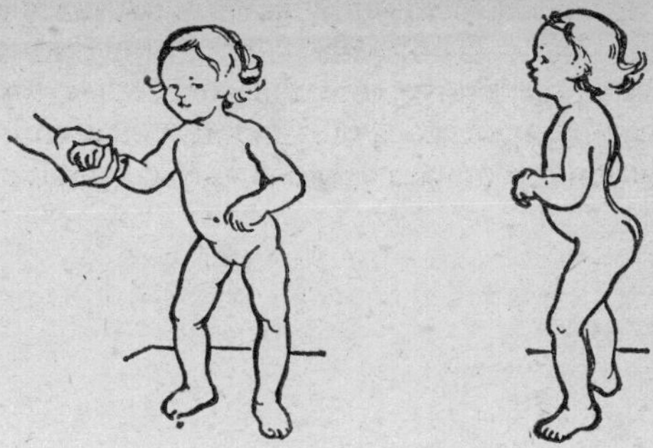

but turn in or out, or if he has an exaggerated curve in the small of his back (as illustrated), there is some abnormality that should be investigated further. These signs may indicate dislocation of both hips.

What Should Be Done. If your baby has difficulty with any of these tests, it may indicate that something is wrong within his musculo-skeletal system. He may have a dislocated hip, a hidden spinal defect, tibial torsion, or a muscle imbalance that is straining his tiny body and starting him on the way to a chain reaction of related abnormalities that can plague him all his life. If you find anything that appears to be wrong, take the child to your doctor and tell him about the tests you performed and what you found. He will tell you whether there is anything to be concerned about and, if so, what you should do.

The important thing to remember always is that the earlier any musculo-skeletal condition is treated, the simpler and easier and less expensive the treatment is, and the better the results. A dislocated hip treated properly early in life may never give the child another day of trouble. If it is not treated until the child is several years of age, it may never heal completely, and the child might be plagued with aches and pains, ineptness and self-consciousness for the rest of his life. If there's ever a doubt, check it out.

In this day of modern medicine, you can almost always

obtain adequate medical care from a physician or a local hospital. Should the physician advise you that there is no dislocation of the hip, go over the exercise programme with the physician. Often you will find that the condition will seem to clear up completely following corrective exercises. Use this only as a *temporary measure*. It is best to visit the doctor again so he can supply other necessary treatment as then seems indicated. Then you will know you are doing everything you can to give your child the best care and happiest future possible.

What Should Not Be Done. There are several things you should *not* do. It's amazing the number of things that some people – both in the medical profession and out of it – can think up to do that in the end cause more harm than good.

I can remember one experience I had quite a number of years ago. A large number of babies were being brought to me with dislocation of the hip. After a while I realized that almost all of them had been born in the same hospital. Why was one hospital involved to such an extent? Was it possible that an obstetrician or paediatrician on that hospital's staff was somehow unwittingly causing all of the dislocations I was seeing? I checked the records further, but the deliveries had been performed by several different obstetricians, so that couldn't have been the problem. And the cases had been referred to me by several different paediatricians. I checked with other orthopaedic physicians in the city to see if the origin of this problem might not be more complex than I expected. But my investigation showed that the number of hip dislocations coming to me was far higher than the number being seen in other sections of the city.

Then one day I was called into that hospital to examine and treat a newborn infant who had club foot. I was in the newborn nursery applying a plaster cast on the blue-eyed little baby. While I was holding the tiny foot in the corrected position, waiting for the plaster to harden, I glanced around the nursery at the always fascinating faces of the red and wrinkled new arrivals to the world. I did a double take when

I saw one baby lying face down with sandbags on top of his bottom. What in the world was this treatment? I had never seen or heard anything like it.

I went to the nurse in charge and asked what type of treatment this was and what condition it was being used for. The nurse said she had decided to try to speed up the straightening of the babies' hips with the sandbags so that they would be straightened out by the time the mothers left the hospital and took their babies home.

A very short time after that she was relieved of her duties at the hospital nursery, and I was relieved of a long list of cases of dislocation of the hips in the newborn.

There are common practices that well-meaning adults inflict on helpless, vulnerable infants that can cause later damage. Among them are holding a baby up by its feet with its head down; or forcing it to stand before it is ready. Any of these can cause dislocation of the hip.

In some cultures it is customary to wrap tightly or swaddle infants from the time they are born until they are ready to walk. Part of the reason for swaddling was to produce an air-tight wrapping, since frequent nappy changes were considered dangerous. The result was a high incidence of dislocated hip, as the swaddling constantly put pressure on the thigh bone and forced the leg bone outward. Similar to swaddling in its effects was the practice of the American Indians of tying their infants into a papoose board at the stage of their development when their soft skeletons could be triggered into dislocation.

On the other hand, some cultural practices discouraged dislocation of the hip. In China and Japan, babies were carried on their mother's backs with their legs in a straddle position, so that congenital dislocation of the hip was virtually unknown. The straddle position was good for cases where dislocation occurred before birth, too, because it helped stretch the leg muscles and ultimately, in some instances, corrected the dislocation. Since this custom has been abandoned there has been a resurgence of hip dislocation in the Far East.

To Summarize. There are some physical tests you can put your infant or child through to determine whether he has muscle imbalance or dislocations. However, usually your own eyes are the best judges. If you notice anything about your child's physical development that worries you, take it up with your doctor promptly.

In the next chapter there are descriptions of some of the most common orthopaedic ailments of childhood, and some of the ortho-exercises I recommend for use in treating them.

CHAPTER SEVEN

Orthopaedic Problems in Children

My youngest patients, like those of any other orthopaedist, have ailments that range from the mild and curable to the heartbreakingly severe. Yet no matter how serious any child's disorder is, it is always with a great deal of hope that I examine him and begin treatment. For over and over again I have had the pleasure of seeing even a seriously handicapped child make unbelievable progress. Hundreds of children have not only achieved full use of reduced, dislocated hips, recovered from fractured limbs, osteochondroses and various forms of malposture but, by sticking to an ortho-exercise schedule, have also corrected the muscle imbalance that led to the dislocation, fracture, osteochondrosis or malposture in the first place. What is more spectacular is the restoration of muscle balance and tone while avoiding the secondary adaptations or complications peculiar to these disorders.

When children need exercises, both parents should visit the therapist so that he can show them the proper way to carry out the necessary exercises. Even if the mother is going to help their child with the exercises, the father should be present, also, so that he can help when necessary, and also check to see that the mother is doing the exercises correctly. The doctor should re-examine the child every seven weeks or so to see how much progress he is making and to see whether some of the exercises can be discarded and new ones added as he progresses.

Growing Pains. Pains or cramps in the legs, feet, hips and spine are often called 'growing pains' and casually dismissed as 'just one of those things that go with growing up'. All too often a youngster and his parents are told that they'll be outgrown.

In reality 'growing pains' are usually the result of muscle imbalance, which evolves during rapid growth and results in distortions of the musculo-skeletal system.

I cannot emphasize often enough that growing pains are not normal. They are a symptom of muscle strain. For my young patients I recommend a programme of exercises to strengthen the hip and leg muscles or spine so that they can keep pace with the demands of the child's rapidly growing skeleton.

Your doctor will pinpoint the site of musculo-skeletal imbalance and recommend appropriate exercises and shoes.

Congenital Wry-neck (Torticollis). In this birth disorder, the neck is twisted and the head is cocked to the side. Often it is only noticeable several days after birth as a lump in the muscle area on the side of the neck. Often the physician will relate it to difficulty with the mother's labour. It is usually not the difficult delivery that produced the lump but rather, the contracted neck because of muscle imbalance in the foetus that is the cause of the difficult delivery.

It may not be discovered until about age three or four, when the tilting of the head becomes noticeable. When this happens, the ear on the lowered side seems to become shorter and wider and to stand out from the head. If the child is lucky enough to have had someone notice the lump on the neck muscle early, exercises can be started even in infancy to prevent the development of wry-neck symptoms and resultant facial deformities.

Only a physician can diagnose this ailment, and no parents should attempt to treat the condition without direct medical supervision.

Congenital Dislocation of the Hip. It is no one's fault when a child is born with a dislocation of the hip and the mother and father should have no guilt feeling about it.

Dislocation of the hip in a newborn infant has been known since the days of Hippocrates, but it came as an entirely new and frightening concept to Mary and Sam Perlman when their baby girl was born with this problem. She was perfectly normal otherwise. Their paediatrician explained that the hip dislocation had occurred in the uterus before the baby was born, and that it was not the fault of either parent. At one point in its pre-natal existence, the foetus had changed its position in such a way that the head of the baby's leg bone was pulled out of the hip socket. The hip joint was particularly susceptible to stress because as the foetus developed the muscle growth did not keep up with the bone growth. Due to the muscle imbalance the foetus's hips did not rotate properly in the uterus.

Fortunately, the Perlmans' doctor recognized the hip dislocation right away, and he was able to assure the anxious parents that the problem could be reversed. The mother was taught exercises that she could use to stretch and strengthen her infant's hip muscles. Before her first birthday, the child's legs had a normal range of movement.

The exercise programme was continued even after the legs developed normal range of movement. Too often I have found that between the ages of five and twelve years, the abnormal pull of the iliopsoas recurs at the hip or at the spine. The presence of a shortened iliopsoas during the rapid growth spurts of the child results in 'scoliosis' of the spine. Therefore, the parent should be alert for hip tension signs in the child with congenital hip dislocation until he is in his teens.

The Perlman baby was much luckier than an older child I saw recently, whose dislocated hip had not been recognized at birth. As she grew older and started to walk, mortar and pestle action of the leg bone in the hip socket slowly wore away part of the head of the leg bone. As she gained weight and walked more and more, the increased weight and pressure on the weak hip joint made her develop a sway-back and a strange way of flinging out her feet when she walked. This poor child had to wear a series of casts for about four months, and then had to wear a splint at night for almost three years. With these aids, and an intensive exercise programme, we were able to undo much of the damage eventually.

Dislocation of the hip occurs ten times more frequently in girls than in boys because the wider female pelvis is more vulnerable to dislocation.

The first step in treatment is known as reduction. The physician bends the affected leg up and, exerting gradual pressure on the leg with his open palm, gently kneads and rotates it until the leg slips into place with a snapping sound that can be heard from across the room. Usually a cast is put on to hold the hip securely in this position while the muscles and tendons rest and grow. Depending on the age and growth-rate of the infant, the cast may be changed every four to six weeks. After the cast is removed, muscle stretching exercises must be done daily. The doctor will teach the parent or nurse exactly what is to be done.

In the exercise I usually recommend, the infant is placed on his back on a table with his good leg bent up in the foetal position. Then the recovering leg is very, very gently pressed down with brief, repetitive movements of the parent's hand on his thigh. This should be repeated for only a few minutes at a time, two or three times daily. There should *never* be rapid extension or hard pushing; force must *never* be used because it can produce a severe dislocation in the opposite direction. The exercise should be done throughout infancy and, if so advised by the physician, into early childhood, to be certain that the supporting muscles have been strengthened, and are in balance.

***Congenital Bent Hip* (*Coxa Vera*).** The condition known as Bent Hip or Coxa Vera is a variation of congenital dislocation of the hip, which is usually seen in young children. In this case the head of the thigh bone slips forward in the hip socket, instead of out at an angle, as its normal anatomical position would be. The patient will seem to have bow-legs and knock-knees on one or both sides. A child may complain of stiffness, weakness or pain after strenuous exercise, or he may have muscle spasms in the hip or in the knee. In toddlers and older children there is a noticeable limp; and if the disorder occurs in both hips, the child has a distinct, waddling type of walk with obvious oscillation of the pelvis. If he walks slowly, the pelvis and trunk seem normal, but when he walks fast or begins to run, the pelvis and trunk sway very exaggeratedly. Occasionally coxa vera appears in only one hip. It may cause a shortening of the leg or knock-knee on the affected side. Often the child can't turn his thigh inward.

The treatment is basically the same as for dislocated hip (although in this case the head of the bone has not left the hip socket) – long-term exercise therapy, always under medical supervision.

In older children the hip ailment most frequently seen is a form of sprain caused by excessive stresses placed upon hip muscles that are still too immature to withstand them. Sometimes an accident may bring the condition on, but just as often it is routine wear-and-tear that ultimately places an unbearable burden on imbalanced muscles.

Coxa Plana; Leggs-Perthe's Disorder. This condition occurs commonly between the ages of four and ten years, more often in boys than in girls, and usually on just one side. It is one of the most common causes of painful limp in childhood, and needs to be treated promptly because it can lead to severe disability in adult life.

The child will complain about aching in the knee and thigh. He avoids running and occasionally limps. The child will usually turn his toes out when he walks; and he may have the duck-waddle type of gait seen in other hip disorders. One leg may be knock-kneed and shorter than the other. If untreated, the muscles will weaken further and there will eventually be noticeable X-ray findings of bone destruction at the hip joint.

Treatment usually consists of a combination of bracing and strengthening exercises to correct the disturbed body mechanics. The exercises will strengthen some muscles and lengthen and relax others to re-establish the normal hip joint alignment of the head of the thigh bone. In addition to the exercises, the child should wear a non-weight-bearing brace for a while and his shoe should be elevated on the normal side about two to four inches. Your doctor will inform you if the exercises are not effective enough in releasing the contracted muscles; surgery may then be necessary.

Lenny Grantley was eight years old when he came into the office. He had a bad limp. X-rays showed that the head of the thigh bone, where it fits into the hip, was flattened away by wear, and the blood supply to the head of the bone was decreased dangerously. Lenny had already been to other doctors. One had told his parents he had to have complete rest in bed and traction for several months. So his leg was tied up and held in place with weights and pulleys, and he and his family shared the problems of keeping a normally active child in one small place for what seemed endless weeks. But after the traction treatment, his symptoms seemed to recur, so he was taken to another doctor, who used braces and crutches to rest the weak leg. By this time Lenny had been undergoing treatment for two years, a long time for a kid not to be out playing with the gang. He was a pretty discouraged boy by the time he came to us and began exercises to stretch the leg and hip muscles and take the pressure off the troubled hip. After following a very carefully planned exercise programme for several months, he was out playing with his friends again. We keep checking on his condition now with X-rays every year, and he must stay with a moderate exercise programme. If signs of stress start showing up again in the hip, he will have to return to a more intensive exercise schedule. We don't want him to have arthritis in that hip when he gets older.

If you suspect that your child has this condition, see your doctor. Exercises should only be done under medical supervision.

Slipped Upper Femoral Epiphysis. A problem similar to coxa plana that occurs during the second growth spurt of the teen years is known as slipped upper femoral epiphysis. It occurs when there is a discrepancy between the growth rates of the muscles and the bones, causing a strain on the hipjoint. Finally, the pull of the muscles makes the upper part of the thigh bone slip out of place. Because the bone is more mature, it does not crumble as it does with coxa plana.

This is what happened to Al Blake, one of those addle-brained adolescents who can be quite charming. You couldn't help loving him. He was a handsome lad who had a rapid spurt in growth at the age of fifteen. Overnight, the baby-faced small boy became a tall, lean, good-looking fellow. But in the course of that rapid growth, his bones had developed somewhat more rapidly than his muscles, and the muscles were now pulling and straining at his hip joint. This caused the pain, soreness and limping that brought Al to my office.

X-rays showed that the muscle strain had already caused some slipping of the neck of the thigh bone away from the head. Luckily the bone had not slipped completely, as it sometimes does in such cases when a sudden additional strain is placed on the already weakened tissue.

This disorder appears frequently in young people between the ages of eleven and fifteen, when the teenage growth spurt occurs. Usually the hip disorder is seen in overweight teen-agers who often also have poor posture, weak arches and knock-knees, and who are clumsy in physical activity. However, the same problem can occur in a tall, thin child if he experiences a rapid enough growth spurt.

Without prompt treatment, the lesion slowly and steadily progresses, while fibrous scar tissue gradually covers the head of the thigh bone. Eventually the degeneration of muscles and bones will have a chain-reaction effect on the small blood vessels supplying the area; and this impairment in the blood circulation can, in its turn, cause still further degenerative changes.

When I saw Al, I asked how long the problem had been going on. His mother said he had been limping for about two months and in the last few weeks had felt an increasing stiffness and limitation of motion in his hip. He had started to have pain in the hip whenever he did much walking or athletics, and occasionally felt a stab of pain at the inside of his knee. But he, like most young people, was 'too busy' and didn't want to be 'bothered' by a visit to the doctor. Finally, the pain became severe enough, and hindered his athletic and social life enough, for him to give in to his mother.

It was necessary to put a nail in Al's bone to strengthen it. Had dynamic testing been performed on Al when he was a young child and corrective exercises prescribed, this condition could have been prevented.

However, even with a nailed hip until Al's bone growth period was over, he would remain vulnerable to the same problem. Therefore I put him on a programme of stretching exercises to help his muscles 'keep up' with his bones. Otherwise, he might have had hip trouble for the rest of his life.

As you can see, hip pains can indicate serious problems. Whenever you suspect a hip disorder in your child, see your doctor. Only he can interpret the symptoms with the accuracy needed for an effective diagnosis. And only he can prescribe the specific treatment needed for each case.

Round Back. Round back and round shoulders are chain-reaction results of rigid hip muscles, which tilt the pelvis forward and force the upper back (dorsal spine) into a 'rainbow' arc. This malposture is common in adolescence.

The child sits, stands and walks in a slumped manner and frequently complains of vague pains ('growing-pains') in the dorsal spine. Usually the parents constantly prod the child into standing or sitting erect; a posture which he cannot assume because of muscle imbalance.

Typically, the teenager with this condition has round shoulders and a 'sway-back'. Often he has flat feet as well.

It will take more than will-power to correct posture problems such as these. They are symptoms of muscle imbalance, and only strengthening by exercise of the spinal, leg and hip muscles will get at the root of the problem.

Check with your doctor about doing exercises.

Start with fifteen minutes twice daily of ortho-exercises, 1, 2, 3, 4 and 10 in Chapter IV, doing them slowly 100 times each. At weekly intervals, speed up your rate on these ortho-exercises. By the fourth week you should be able to see and feel improvement. A firm mattress is recommended.

Spondylolisthesis. This jaw-breaker term describes a condition, sometimes occurring in children of age three or four, but more frequently seen during the rapid growth period of the child between five and eight.

Pressured by musculo-skeletal imbalance in which the iliopsoas, quadratus lumborum (an important muscle which goes from the pelvis to the bottom ribs) and hamstring muscles are the prime culprits, one of the vertebrae in the lower back may slide partway off its cushioning disc until it rests on the vertebra beneath it. The vertebra may slide only slightly out of position, or it may be totally displaced.

The back muscles often tense up to withstand the strain, and may become spastically rigid. The slipped vertebra may be experienced as little more than vague 'growing pains', or as severe pain in both the back and legs, leading to a peculiar way of walking.

If your child has any of these symptoms, see your doctor. One diagnostic test is whether the child has difficulty raising his leg without bending his knee.

Usually, support in the weakened and strained spinal area over a period of time, using a surgical corset or brace, will give the displaced vertebra a chance to stabilize. Total rest in bed and aspirin will help the child get through a bad attack. In addition to these measures, I recommend a continued programme of exercise therapy to strengthen the muscles of the back and pelvis and thus prevent further damage (especially low back pain exercises; see pages 87–92). Your doctor is the only one who can properly programme your treatment. He may be conservative at first but, if the treatment is not successful, he may recommend surgery.

Scoliosis. No other deformity presents such a magnitude of intricate and baffling problems to the physician as does scoliosis – curvature of the spine. Recognition of this deformity can be traced as far back as the time of Hippocrates, who first applied the term 'scoliosis'.

Among the many cases that I have treated, that of Linda Liebman could be considered typical. I first saw her when she was ten years old, and the sideways curve of her spine was quite noticeable. Her mother said it had appeared gradually at first and then progressed remarkably in a very short time. This is a typical pattern because scoliosis develops in the growing child and worsens as the growth of the spine continues. Usually there isn't any pain, but the spine becomes more and more curved as the child grows older. In the beginning the parents may not even notice the curvature; most often it starts to show up just when the youngster has learned to bathe himself. At first one hip may appear a little higher than the other. Then the ribs and shoulder girdles start to curve out of line as the child's growth continues. The curve of the spine becomes progressively more and more pronounced. When spinal growth stops at the age of seventeen or eighteen the deformity, if it has not been corrected, remains for the rest of a person's life. In an adult the condition can be very severe and disabling.

What is quite puzzling to a mother is that one doctor will prescribe a Milwaukee brace for her daughter, another a Risser plaster jacket, another a Cotrel plaster jacket, another head and pelvic traction, etc. Remember each doctor will prescribe that form of treatment for each stage of scoliosis which in his own experience has produced his best results. What is a mother to do? Rest assured that there is one form of treatment on which all doctors will agree, and that is active exercises.

Scoliosis also provides one of the most dramatic examples

of what orthotherapy can accomplish, especially in the early stages of spine curvature before the bones of the spine and ribs become fixed. The treatment is based on strengthening the weakened muscle groups by selected exercises, which at the same time help to stabilize further progression of the curve. Your doctor can prescribe and regulate your exercise programme. If need be he will also prescribe the use of a plaster cast or some kind of brace. As a last resort, surgery may be necessary. But, as in virtually every orthopaedic problem, the earlier the condition is treated, the less radical the treatment and the better the results. Therefore the best form of therapy is prevention. Through dynamic spine-pelvis-hip testing, especially during the period of rapid growth, the earliest signs of muscle imbalance can be detected, and these pattern a programme of ortho-exercises for correction.

We had Linda do a series of exercises specially designed to strengthen and increase the flexibility of the muscles of her back, hips and legs, as well as some of the regular ortho-exercises to relax the muscles that were pulling her spine to one side. Our aim was to relieve the strain on the joints within the spinal column. Usually the iliopsoas and the quadratus lumborum trigger the spinal disorder. However, prior to the development of the curvature the kinetic chain reaction of muscle imbalance also involves the muscles of the buttocks, hip, thigh and calf. Therefore, in order to reach the iliopsoas, the cycle must be reversed; that is to say that the calf muscle imbalance must first be corrected, then the thigh muscle, then the buttocks, the hips, and finally the iliopsoas and another important muscle in the back, the quadratus lumborum.

Because no one should attempt to exercise a curvature unless under direct medical supervision, we did not include the special exercises for these cases in Chapter IV. The special scoliosis exercises are described on pages 96 to 104. I also use ortho-exercises 4 and 10. If you suspect that your child has scoliosis consult your physician before you begin any exercise programme.

Clubfeet. A clubfoot is any deformity in which the foot is twisted out of shape or position. It can take many forms. Sometimes a child walks on the outside rim of the foot, sometimes on the heels with the toes drawn up, sometimes on the toes. The condition is almost always due to a congenital birth defect. It can also be caused by poliomyelitis or various muscle disorders.

Most cases can be treated successfully, although about one out of five relapse and then need further treatment. The earlier the beginning of treatment, the better the results. Even a slight deformity in a newborn infant should be treated. It is important to carry out exercises and other treatment for as long a time as your physician recommends. Don't stop just because the foot seems to you to be corrected. If not treated, the condition will gradually get worse, especially when the child starts walking and puts weight on the already weak foot and ankle.

First, the alignment of the foot is corrected through periodic manipulations by a physician, gently forcing the foot little by little towards as much correction as possible. Then a cast, a splint or adhesive strapping is used to maintain correction. Sometimes surgery is needed. Every day the mother should do stretching exercises on the child's foot. After he learns to walk, the child should have corrective shoes.

Claw foot and hammer toes are conditions similar to club foot, consisting of excessive curvature of the sole of the foot and toes. Corns and callouses often add to the discomfort, tenderness and pain on the ball of the foot.

To relax foot muscles I recommend the special exercises for sore feet (see page 116). Adjusted corrective footwear is also recommended.

Bow-legs. Bow-legs or 'bandy-legs' is an outward curving of the knees or the leg bones. Bow-leggedness at one time was associated primarily with rickets, a vitamin deficiency disease affecting the bones. Today rickets is rare. Vitamin D and viosterol have done much to eliminate the disease.

Yet, bow-leggedness is still very common. With the advent of the mini-skirt, a slow walk down any street will reveal the high percentage of bow-leggedness to the sharp-eyed observer. I used to think the resurgence of this condition was peculiar to the United States. But when I walked down the avenues of the capitals of Europe, Northern Africa, the Middle East, Australia and the Far East, especially in Japan, I discovered that bow-leggedness has taken hold all over the world.

Unless the bowing is advanced, it starts by producing symptoms usually considered 'growing pains'. The real culprit is the tibial torsion syndrome, which is placing extra stress on the foot, knee and hip. As the child grows and the condition worsens, he may develop 'flat feet', and his lower legs will start to turn in at the knee. Finally, the growth of the inner side of the upper part of the leg is depressed, while the outside continues to grow – actually curving the leg bones out.

Usually, during the very early stages, the parent can feel a door-knob-like projection by pressing along the upper and inner part of the leg just below the knee cap. During the early years of growth, these projections become sensitive to pressure. Another way to determine the extent of bow-leggedness in your child is to check the distance between the child's knees when he is standing up straight with his feet together.

By the time a child has become aware of his cosmetic disfigurement, it is usually too late for conservative measures. I would like to see every newborn routinely checked for tibial torsion syndrome, so that this could always be corrected early in life, long before adaptations occur. Check with your doctor if you suspect that your child is developing bow-leggedness. He will suggest the proper correction in his shoes, (see page 149) and will guide you in an exercise programme.

The bow-legged child will be helped by any one of the leg and foot exercises already described, as well as by ortho-exercises 14, 16, 17, 18 and 19.

Knock-knees. Another childhood complaint that is so frequent as to be falsely dismissed as 'normal' is knock-knees. This, too, will *not* be outgrown, and is in fact a symptom of a more complex problem, tibial torsion syndrome.

The bending together of the knees results from the natural attempt of the muscles of the lower leg to compensate for imbalance in the musculature of the hip joint and upper leg. This condition may become noticeable at any time after the child starts to walk; usually it becomes apparent when he is around three years old, but sometimes it is not noticed until the rapid growth spurt of the teens.

Improper use of scaphoid pads or wedges in the shoes can also cause knock-knees. This does not mean that you should not use these supports if there is a need for them, but it does mean you should be watchful. Wedges on the inner or outer side of the heel and sole may precipitate the development of knock-knees in the thirty per-cent-ers.

If you notice your child standing with his legs apart and pushed backward at the knees, shoulders rounded and stomach out, he is probably on his way to developing knock-knees. These signs of poor posture are also symptoms of tibial torsion syndrome adaptations or compensations. If shoe wedges are the problem, of course remove them. An exercise programme to correct knock-knees starts by working on the ultimate cause, the tibial torsion syndrome. The child should do ortho-exercises 14, 16, 17, 18 and 19, as well as the special leg and foot exercises (see pages 113 and 117).

In the child with rickets, poor nutrition is causing the bone and muscle malformations, and all the exercise in the world will not help unless the child's diet is corrected and corrective footwear worn. When proper nutrients are made available to the growing bone and muscle tissue, then ortho-exercises can work on these tissues to correct the damage that has already been done (in extreme cases, surgery as well as improved diet and corrective shoes may be required).

Recurrent Dislocation of the Knee. The term 'dislocation of the knee' is usually used to describe the accidental movement of the patella, or knee-cap, to the outer side of the knee joint. This usually happens while the knee is bent and the foot is turned outwards. In extreme cases, an injury may not only effect dislocation, but also sustain dislocation. The knee cannot be straightened. Sometimes the doctor can manipulate the leg so that the knee-cap snaps back into place. Often the knee-cap will snap back into place by itself within seconds after the injury takes place.

I have seen many of these dislocations. Many of them were precipitated when the patient suddenly fell forward while holding his foot outward and bending his knees, or by automobile accidents in which the knee strikes against the dashboard. I say precipitated, not caused, because more often than not the person who dislocates his knee is a thirty per-cent-er, predisposed to this condition by a musculo-skeletal imbalance of which he is unaware.

I recently had an opportunity to examine three generations of the same family. They had come to me because three of the five daughters had recurrent dislocations of the knee-cap. Their mother had had this condition since childhood. One of the affected daughters had just had a baby who had a congenital dislocation of both hips. This family had inherited and passed on a tibial torsion syndrome which later tended to reveal itself in the form of a dislocated knee cap.

The usual sufferer of dislocation of the knee is an average-weight female with knock-knees. Extra weight will emphasize the problem, since heavy thighs tend to push the knee cap even farther to the side. When a patient stands in front of me, I look to see if her feet turn outwards. Even when she stands up straight, with her knees pointing straight ahead, her ankle joints are not directly under her knees. The twist in her leg muscles is exerting constant pressure on the knee cap. Any sudden bending of the leg will tend to jar the knee cap off to the side. It may snap back the first few times. But if it is dislocated enough times, all the muscles around the patella will become imbalanced and weaken until the condition becomes a chronic one termed 'habitual dislocation of the patella'.

If you suspect this condition, visit your doctor, as he alone can make the diagnosis and outline treatment. Many surgical procedures have been devised to correct this condition, but each demands some sacrifice of motion in the knee. Although I have operated on many of these disorders, I would prefer to see my patients correct the basic muscle imbalance which caused them.

I recommend the special exercises for torn ligaments of the knee (see pages 109–10) and ortho-exercises 14, 16, 17, 18 and 19 to correct the postural imbalance which is the source of the problem.

***Inflammation of the Knee Joint* (*Osgood Schlatter's Disorder*).** Simple inflammation and pain in the knee often occurs in boys between the ages of ten and fifteen years. It can be caused by a sudden pull on a leg muscle (rectus femoris), by a direct blow on the knee, or by the constant pulling of tight leg muscles. The child complains of pain when running, kneeling, climbing stairs or bicycle riding. There may also be swelling, and the child may limp.

To relieve the immediate discomfort, a heating pad or hot bath are helpful. The child should take aspirin or another mild pain reliever. The pain will gradually subside over a period of days or weeks. When the pain stops, the child can gradually resume his normal activities. However, if symptoms are severe and persistent, your doctor may recommend a cast or other form of management.

An exercise programme to strengthen knee muscles in youngsters should begin with ortho-exercise 21 and proceed through the special knee exercises (see pages 109–10).

Pump Bumps. A painful callus at the back of the heel is common in sub-teen and teenage girls. It usually occurs when girls with tibial torsion syndrome wear a pump-type shoe with a narrow heel and an arch on an inclined plane. The collar of the shoe heel rubs up and down over the heel bone, causing a painful bump. When the girl changes to a more stable shoe with a wide heel, the symptoms subside. The elimination of the friction between the heel counter and heel bone will usually correct the condition.

Pigeon Toes. As soon as you notice that your child's toes turn in, exercises can be started, even if it is in the first few months after birth. Gentle manipulative exercises can stretch the tissues and work the foot back to a normal position. In extreme cases, your doctor may prescribe a series of corrective casts, with casts being changed every few weeks as the foot grows and begins to become normal. The doctor may prescribe a corrective boot to be worn until the child begins to walk; or straight, last-supportive shoes with a heel pad.

A pigeon toe condition that is not corrected can lead to secondary effects later in life, including constant tiredness and sore feet. As the big toe turns outward, it may form a hammer toe, or the bony joint on the toe may form a callus and a bunion, causing greater discomfort and disability.

The doctor may prescribe exercises to strengthen the foot muscles, such as those already described for sore feet (see page 116), and corrective footwear.

Shoes for Infants and Children. The basic requirements for shoes during the child's growth period are adequate length and width. The sole of the shoe should be flexible and the leather soft. When the child is standing, the big toe should be a thumb's width from the end of the shoe. Use the pinch test to determine the proper width: leather over the widest part of the foot should be loose enough for a tiny bit to be pinched up.

For the child without musculo-skeletal imbalance, almost any style of shoe which meets these requirements will give adequate support. The child with musculo-skeletal imbalance, especially the child with tibial torsion syndrome, needs more from his shoes. This child's shoe really must be made of leather, for leather 'breathes', while man-made products do not. A steel plate running from the heel of the shoe towards the sole is needed to support correctly the arch of the foot. His shoe should have a built-in extended counter, what is called a Thomas heel, and about a quarter-inch heel pad.

Buying shoes for the thirty per-cent-er can be an exhausting process. You may find yourself taking your child from store to store. Never allow a shoe salesman to talk you into allowing him to insert wedges into your child's shoe that have not been specifically prescribed by a physician.

I have a terrible time prescribing shoes for my patients and for my children. Every shoe salesman knows precisely what I am talking about, but with the exception of a few salesmen, I have rarely been satisfied. It is not the salesman's fault. The shoe manufacturers are not supplying him with adequate shoes for the thirty per-cent-ers.

I have been travelling to Washington for years, trying to convince a government regulatory agency that measures should be taken requiring shoe manufacturers to produce the kind of shoe that the thirty per-cent-er needs. With today's styles in shoes, I see no reason why shoes could not be manufactured which are corrective, supportive and fashionable.

Today's parent is forced to drag his child to the orthopaedic shoe shop where he pays a high price for a shoe his child is embarrassed to wear. Often, the child simply refuses to wear his corrective shoes. Even more often, a mother will not even realize that her child needs a special shoe. She will go to a reputable shoe shop and buy what seems to her to be a sturdy shoe. The salesman will tell her that it fits well, and she will leave the store, convinced that she has done the best for her child. In fact, that child may be missing the golden opportunity to correct his muscle imbalance. Wouldn't it be wonderful if the shoe salesman could be trained and equipped to recognize the thirty per-cent-er, and supply him with the kind of shoes he needs?

The child with any muscle imbalance should not go barefoot, or wear sneakers or other soft shoes, on concrete, wood or other hard surfaces. Theoretically, it would even be possible for shoe manufacturers to make sneakers with the proper support, but so far it has not been done. I am afraid that only concerted pressure from the public will reform the shoe manufacturing standards in this country.

Orthopaedic Devices. Casts are very effective in helping a baby with tibial torsion if applied before the child is ready to walk. The casts keep the legs in the proper position and prevent strain on the joints from the unaccustomed weight-bearing as the child learns to walk. A child can learn to walk without difficulty while wearing the casts, which are usually left on for about four weeks before being either replaced by another set or removed entirely.

Among other frequently used devices are the de-torsion apparatus to correct improper rotation of the bones; the night splint used to correct the direction of abnormal muscle pull; the Milwaukee brace, extending from the chin to the waist, which may have to be worn twenty-four hours a day to exert a constant firm pressure on the neck, shoulders and spine; and other types of braces for special conditions. Orthopaedic girdles and heavy elastic rib belts can also be helpful in some cases.

A good example of a specialized device is the recently developed 'Clapper-Dapper', intended for children who have foot and leg deformities resulting from cerebral palsy. One little boy, after surgery and stretching exercises, still had fairly severe spastic leg paralysis. But he absolutely refused to wear the braces prescribed for him, violently screaming and ripping them off, until his parents had to give up. The child's grandfather, an engineer, knew that the purpose of braces was to hold the child's legs in proper walking position, and particularly to prevent him from rising onto his toes. The grandfather designed a kind of short metal ski which could clamp on to the sole of the child's shoe. It kept his legs in the correct position and improved his gait tremendously.

The child's doctor, Dr David MacFarlane of the University of Oregon Medical School in Portland, was so impressed that, after modifying the design somewhat, he tested it on more than fifty young patients, with excellent results. The name of this device came from the clip-clapping noise made when the child walks, which can easily be eliminated by putting a rubber rain shoe over the metal.

There are times when a condition is so severe that surgery must be depended upon to alter a given muscle or bone. The results can be quite dramatic. When a child with cerebral palsy, whose arm is turned outward in a torque position, has the contracted muscles that are causing the problem loosened, his hands and arms will regain normal function again.

In conjunction with all of these treatments, the person should also do corrective exercises. This is the only way to be sure that the muscles continue to do their work properly.

CHAPTER EIGHT

In the Beginning: Causes of Muscle Imbalance

One day a nine-year-old boy was brought into my office by his mother. It was his poor posture which, contrary to the opinion of the paediatrician, had not corrected itself but, indeed, seemed to be getting worse, that led them to consult me. When I took his medical history, he complained of frequent pains in his legs and hips, and that he felt restless when he had to sit and study because he simply couldn't relax in a comfortable position. I asked him if he liked sports. He stammered when he told me that he was always the last one chosen for any teams because he was so completely clumsy and inept. It was not surprising that a physical examination showed him to have a curvature of the upper spine. He was a perfect example of physical unfitness in a boy. And, unless helped, he would grow up to be an equally perfect example of physical uufitness in a man.

What makes a person physically unfit? For this boy, as for so many others, muscle imbalance was the basic cause. His muscles were weak here, tight there, causing tensions and strains that were in turn forced onto joints and bones throughout the body, keeping the body from functioning the way that it should.

But what causes muscle imbalance? Why do some people have muscle disorders and others do not?

There are many influences on muscle development and many points at which things can go wrong. The disturbances can occur as early as the embryonic stage, when the infant's body is first developing. Or the disturbances can occur in the vulnerable growth periods of childhood and adolescence.

They can occur due to accidents or for other reasons in adulthood. And they may even have their origin way back in the course of human evolution.

Let's start from the very beginning.

How the Muscles and Bones Form. The muscle system starts forming and growing when the tiny human embryo is still in its mother's womb, barely more than a worm bathed in the cushioning and nourishing liquid, growing hour by hour and day by day into the patterns that make up the human body.

The tiny future human starts as the sperm and egg join and form one single cell. Then this cell divides and makes two, and these divide and make four and these divide and make eight, then sixteen, then thirty-two – until the tiny speck within the womb is a hollow ball of dividing cells. Then, like a crease made in a man's hat, the ball folds in upon itself. And where that crease is, the cells start to differentiate. With all the marvels of science, we still don't know exactly why one cell becomes a muscle, another becomes part of the heart, another part of the brain. But somehow one end of that crease starts to swell a bit where the head will be. And a speck appears where the eye will be, and when the embryo is about four weeks old – and less than one quarter of an inch long – a tiny tube-like structure begins to beat – the embryonic heart. At the same time other cells are differentiating – cells that will become part of the muscles and bones. The muscle cells start to differentiate in such a way that they can contract to get short, then elongate again. The baby bone cells absorb various minerals that will later make them hard. Soon actual muscles start to form, and tiny bones.

At about the fourth week after the beginnings of life in the uterus, small swellings or buds appear along the sides of the slowly developing human form. These will become the arms and the legs. Tissue swells into the limb buds, pushing them outward. Soon little indentations appear, dividing one pair of buds into future arms and hands, and another pair into legs and feet.

Joints form where bones meet. In fact, there are tiny

joints in the embryo before it is even three months old. They can be joints which permit little or no movement, or they can be joints that are freely movable.

There is one particularly interesting fact about the development of the unborn human. The earliest embryonic stages of development have a definite resemblance to the stages of human evolutionary development over the course of hundreds of thousands of years.

This concept is expressed in a catchy phrase to which every student of embryology, and most students of biology, are exposed. It is a very significant three-word phrase: 'Ontogeny recapitulates phylogeny.'

'Ontogeny' – the development of the fertilized egg; 'recapitulates' – repeats; 'phylogeny' – the evolutionary history of the species. *Ontogeny recapitulates phylogeny* simply means that in the course of their development the structures of the embryo briefly resemble related structures of animals in earlier stages of evolution. The spinal cord, for instance, resembles in its earliest development the soft, simple spinal cord of the larva-like animal called an amphioxus, and only later begins to look like a human spinal cord. The developing heart of the human embryo begins as a one-chamber pulsating tube just as it is in lower animals; then it develops into the more complicated heart we know. The spine of the human embryo even has a 'tail' which is reabsorbed and disappears.

The spine starts forming during the first month, and hardening when the embryo is six to eight weeks old. But the hardening, just like that of the long bones, goes on throughout childhood and into adolescent life.

The muscles, too, develop early in the fifth or sixth week of pre-natal life. By the end of the fourth month, they are able to produce movement – movement which, as any mother knows, can become quite strong before the baby is ready to be born. At this time muscle imbalance may originate, for it is now that leg and spinal rotation begins. Although we do not know for certain, there is enough evidence to indicate that some positions that are taken by the growing foetus place a strain on various developing muscles and joints. For example, the baby crosses its legs in the uterus. In order to cross, the legs must first twist outward and then lift up. It may be precisely in this crossing-and-rotating-of-the-legs

stage that it is determined whether the baby is to be born as a person with muscle imbalance.

Of all creatures, only man crosses his legs *in utero*. If there is inadequate turning of the lower limbs during the ninth to tenth weeks of embryonic life and the legs fail to cross over, then a deformity of the feet known as *talipes calcaneovalgus* occurs. The newborn's feet are bent upward. However, when one or both legs overcross – that is, cross too far – then *tibial torsion syndrome* results. A severe form of tibial torsion syndrome is clubfoot or *talipes equinovarus*.

Is Heredity Involved? We are not sure how important hereditary influences are in causing muscle problems. Although the various conditions of muscle imbalance are as a rule *not* directly inherited, there does seem to be some hereditary pattern involved. This does not mean that any specific symptom is itself inherited, or even that the weakness of the muscle system is transmitted. What it means is that the *tendency* towards a certain type of body structure may be inherited.

There also sometimes seems to be a hereditary pattern in the tendency that some children have for the bones to develop faster than the muscles, causing stress, strain and abnormal muscle action. Other hereditary patterns occur, too. For example, the shortening of the iliopsoas muscle in the newborn occurs much more often in females than in males. But when males have this problem, they usually have it more severely.

The Crucial Years. Many cases of muscle imbalance seem to originate during childhood and adolescence.

At birth all the muscles and bones of the adult human are present, but they still have much growing to do. The long bones, such as those of the legs and arms, grow by forming new cartilage where the shaft of the bone meets the head. As this cartilage hardens to bone and new cartilage is formed, the bone becomes longer and longer. Finally, in the adult the long shaft and rounded head unite and the entire bone is firmly and permanently hardened.

The curve of the spine changes too. In the embryo the spinal column is shaped like a shallow letter *C*. In the baby the curve starts to change and continues as the child grows, until it resembles an elongated *S* in the adolescent. This is the normal curvature of the spine. It appears in stages. The

first curve develops in the neck region, about when the infant tries to expand his horizon of vision and holds his head more erect. The second curve in the small of the back – the lumbar region – starts to form when the child begins to sit up, develops further when he begins to crawl, and is almost completely developed by the time he stands.

The first seven years of life is the period of most rapid growth. This is a particularly vulnerable period. If the muscles keep up with bone length, there are no problems. During the elevation of the centre pole of a circus tent, if one of the ropes is short it will deflect the pole towards the shortened rope. During the period of most rapid growth, if any of the muscles fails to elongate with rapid skeletal growth, the spine, leg or arm bones will deflect towards the shortened muscle and a cycle of muscle imbalance commences. And if the child has a muscle problem, this is the time to find and correct it – before permanent hardening of the bones takes place, when symptoms become more disruptive and harder to treat.

During these critical first years of skeletal development, and during the growth spurt of adolescence, the skeleton is under much stress and is dangerously vulnerable to development of abnormalities.

From Ape to Modern Man – The Key. It was a momentous occasion when man first rose to his feet and stood erect. The occasion was unheralded, marked by no one, but it was the beginning of human life as we know it today. Erect posture freed man's hands for work and made it possible for him to lift his eyes to the stars and heavens. But standing erect may also have given him much of the back trouble and other posture problems that afflict him today.

There was no one moment at which the ape-like men who were our earliest pre-human ancestors suddenly jumped to the next higher branch of our evolutionary family tree. There were many biological changes involved, changes that occurred gradually and at irregular intervals. Over hundreds of thousands of years, body changes took place in individuals here and there – changes that gave the man who possessed them an advantage over his fellows in coping with a harsh natural environment.

Charles Darwin, whose observations led him to postulate the theory of evolution, used the phrases 'natural selection' and 'survival of the fittest' to describe this process. The more successful individuals were likely to have better nutrition, to live longer and healthier lives, and to pass on their biological variations to more and more offspring. Among these advantageous biological changes were those of improved vision, coordination of movement, increased flexibility of fingers and hands, and – of specific concern to us here – erect posture.

It is believed that our earliest pre-human ancestor had long dangling arms and often dropped down on all fours and scampered along, or squatted with his hands touching the earth. His head was large and heavy, bent forward on a short neck and almost touching his chest. His spine was almost straight. Although his legs could be completely extended, the knees were usually kept partly flexed, not unlike the knees of a chimpanzee.

But as man became more upright, his whole geometry and anatomy changed. Compare the skeletons of Neanderthal man (left) and modern man (right) shown here. Modern

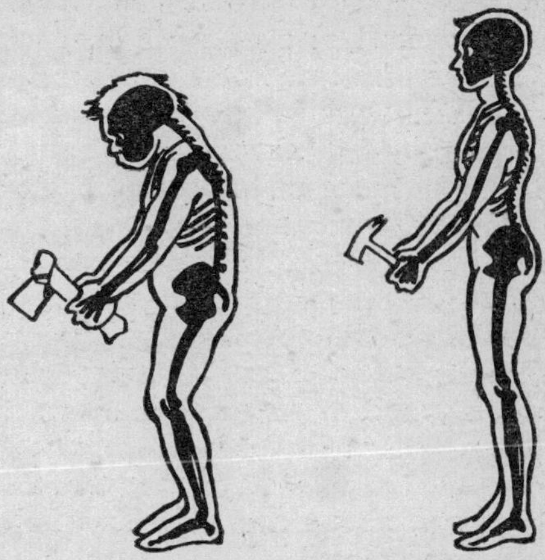

man's head is poised delicately on his straight neck instead of thrusting forward. Eventually his legs became completely straight instead of bent and bowed. His pelvis broadened and the leg bones flared out to support the weight of this upright man. In compensation for all the weight and posture changes, the spine became curved in a shallow S-shape.

Muscles as well as bones underwent adaptations to hold man upright. Muscle attachments shifted to different parts of the bones to compensate for the different movements required for two-legged walking. The calf muscle developed extensively, making it possible for the knee to be fully extended and straightened instead of always flexed as if crouching, and the powerful gluteus maximus – the muscle in your seat – made it possible for the hip joint to straighten out in the same way.

As you might expect, some of the greatest muscle changes took place in the iliopsoas muscle. Over the ages, as man gradually assumed his upright position, the iliopsoas underwent considerable strengthening and elongation. Technically speaking, the points of origin and insertion of the iliopsoas gradually increased, and the muscle became progressively stouter in order to support and stabilize the upright posture.

But the evolution of the iliopsoas has not been uniform in all human beings. For some reason that is only partially understood, many people have an iliopsoas that is too short. It is not elongated and flexible as it is in most people and as it should be. Instead, it has remained short and rigid, pulling against bones and joints and creating constant tension. For individuals who have this defect to maintain an erect posture, some 1,300 pounds of tension pull on their spines and thigh bones. This puts tremendous stress on the hips, constant wear and tear on the back and knees, and eventually can contribute to deformities of the spine, pelvis and hips. It may also secondarily cause muscle imbalance in the legs, ankles and feet. As muscle tension continues and spreads, there can even be permanent injury to bones and joints, or such extremely poor posture as to injure internal organs.

This imbalance of the iliopsoas is a sort of evolutionary 'hangover', in which a third of humanity still does not have this particular advantageous structural change. In a way, it is proof that the evolutionary process is still going on.

We can't help but wonder why this large proportion of the population is, apparently, less 'fit' for 'survival' than the other two-thirds, and yet paradoxically have indeed survived. We can only guess that, in the pre-civilized environment in which man has spent all but one per cent of his years, a short iliopsoas was not a severe hindrance to his survival. Or perhaps many of these symptoms are only showing up now that we walk in constricting shoes on hard concrete surfaces.

But we are concerned now with the quality of life rather than just survival. And if you can get rid of your aches and pains, it may not mean that you will live longer, but it can make you happier and improve the quality of your life.

CHAPTER NINE

Apply the Principles of Orthotherapy to Your Daily Life

The state of your muscles can make the difference between enjoyment or misery, whether you are at home, at work or at school, whether you are riding in a car, playing football or making love.

The state of your muscles determines whether you feel fit and energetic or go through life suffering with aches and pains. It can affect your personality, your emotions, your appearance, your attractiveness to other people, your vigour and stamina, your confidence in yourself and your ability to produce at work.

Most people with muscle problems become so used to living with their miseries, so used to accepting daily pain, that they don't realize they can change their lives. In this book you have seen how pain can be relieved by treating the muscle problems that cause the pain. If you have already embarked on your exercise programme, you know that you can feel better. There are other ways as well of removing unnecessary strain from your muscles. In this chapter we will discuss some of them.

On the Job. Many people are miserable in their jobs because they simply hate to work or hate the jobs they must do. But very often a job is unbearable because of a set of uptight muscles. These people are miserable at work because their backs hurt so that they can't tolerate sitting still at their desks, or standing or walking for the necessary length of time. In fact, the relationship between the body and a specific job is so important to job efficiency and good health that an entire laboratory at New York University Medical Center Institute of Rehabilitation Medicine is now devoted to studying it.

There are many ways in which you can improve your working conditions. Too frequently work tables, chairs and other furnishings are chosen by a company purchaser who does not realize their effect on the muscular comfort and function of the workers. If you discover that the furnishings you must use on your job are putting a strain on your muscles and tiring you too rapidly so that your efficiency is impaired, it is to your company's advantage as well as to yours for you to speak up. The improvement in your productivity when you have proper equipment will benefit you and your employer. In fact, many firms are realizing that poor chairs or desks impair the quality of work, and they have contracted with office furniture suppliers to come to the office and personally fit a chair for each individual.

When I first heard of this, I was reminded of the time, shortly after my wife and I were married, when she insisted that I go with her to buy chairs. Since she had bought all of our other furnishings herself, I couldn't imagine what she had in mind. I was in medical school at the time, and she was working to pay my way. We had both been pre-medical students together, and she must have absorbed more than I about the importance of correct chair fit for comfort and muscle health.

So, at her urging, I went along to sit in the chairs to see if they were the right height. After testing dozens of chairs, I began to realize how right she was. Most of them were noticeably uncomfortable.

All of us spend too many hours of our lives sitting in chairs that are not properly fitted to us. Both in your job and at home, be sure your chair fits you and is comfortable, and don't let anyone else – wife, husband, decorator or salesman – pick your chairs for you.

If you cannot convince your employer to get you a properly fitted chair (give him this chapter to read!), it may be worthwhile for you to bring your own chair to work. It is

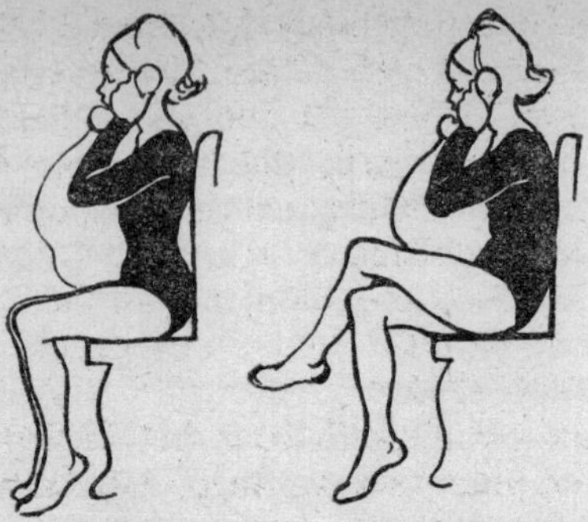

CORRECT INCORRECT

possible to purchase a good chair for a relatively small expenditure; and your saving in work days lost because of back pain or other discomfort may be many times the cost of a good chair.

In choosing a chair, select one that does not cause slouching. This is of particular importance to machine operators, typists and book-keepers. For example, a chair may cause back pain, leg irritation, stiff neck and any number of other ills if it is not the proper fit. The best chair has a low back, a seat that is not too deep, and an arm rest if the nature of your work makes this possible. The height of the seat should be adjustable so that you can find the position most comfortable for you, with your feet placed squarely on the floor.

Another improvement you can make in your own working conditions is to find something to put your feet up on. Whether you work sitting down or standing up, at a table, a desk or a machine, you will be able to relieve a significant amount of muscle strain on your back and hips if you can keep at least one knee bent. Try it and see. You don't need any fancy equipment, either; if there isn't a low stool handy, you can find a sturdy box, a couple of bricks or a cinder block.

You have to analyse your own situation at work. If you are a typist and have suffered from shoulder pains, check to see which of these factors may apply. Perhaps you aren't sitting the proper distance from the typewriter. Your seat may be too high or too low. Your chair might not be supporting your back. The lighting may be poor, so that you have to bend over to read your copy. You may have a vision defect and, in attempting to see clearly, you are putting a strain on your back and shoulders as well as eyes.

The man or woman with muscle imbalance symptoms must pay particular attention to small details. Tools should be chosen carefully. A curved handle usually permits better positioning of the hand and wrist than a straight handle, creating less strain on the arm muscles.

If you work with small materials at a table, the two important elements are the height of thc table and the position of the materials in relation to your body. If the working surface is too low, you have to bend over your work, straining your neck and back muscles. If the surface is too high, you have to raise your hands too high to work, which strains the shoulder muscles. Keep things convenient and orderly so that you can reduce reaching and economize muscle use. These details may not be significant factors for short

CORRECT

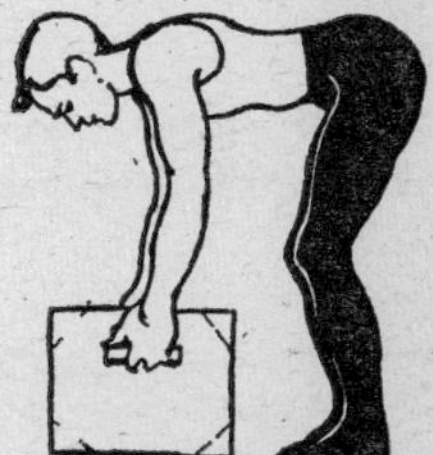

INCORRECT

work periods, but if hour after hour is to be spent in doing one task, they can make a great difference in relieving tension of muscles and general fatigue.

Working positions should be adjusted to avoid your working with your head bent forwards or sideways. Dentists should avoid working in a bent-over position; draftsmen should raise their desks to bring their work closer to them.

If your job involves heavy manual labour, learn the correct way to lift, push and pull. In fact, I believe that everyone should learn this. Who is there who has not at some time needed to move a large piece of furniture or carry a heavy suitcase? Watch professional furniture movers

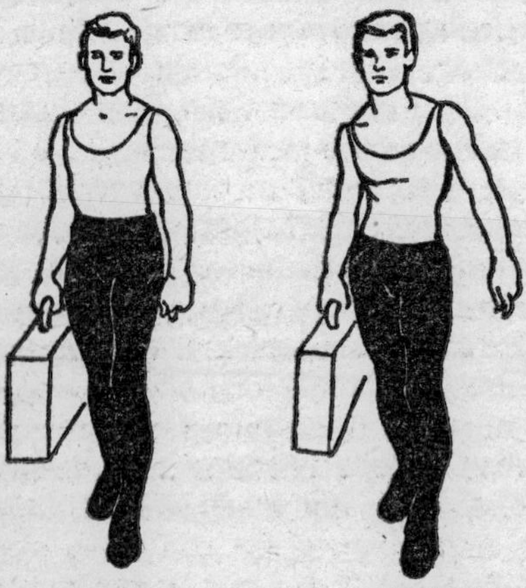

CORRECT INCORRECT

or delivery men at work. You rarely see one of them attempting to move a heavy piece of furniture by himself. They always move slowly, in rhythm, as a well-balanced team. They bend from the knees, hold objects close to their bodies, place their feet far enough apart to provide a strong base of support.

When you must move a heavy object by yourself, keep your back straight and place your feet in such a position that your legs and hips do most of the lifting. Never count on your fragile spine for support. Remember that a light weight held at arm's length from the body produces much more stress on the body than a much heavier weight that is held close in. Also, stress on the body is increased if a woman wears high heels when lifting something.

I have treated many people for pains in the arms and neck when they first start wearing bi-focals, or when they change from bi-focals to tri-focals. The placement of the reading section of the lens at the bottom may cause the wearer to stretch his neck awkwardly if he must read material that is at eye level or higher. The constant stretching of the neck leads to an imbalance of the neck and shoulder muscles.

Most people will note onset of symptoms within two to four weeks of obtaining the new lenses. Among my patients with this problem have been an ophthalmologist who customarily sat while doing refractions, a short radiologist who had to study films on a tall viewing box, an aeroplane pilot who had difficulty viewing his instrument panel, a scientist who used a microscope at a high table, and a housewife who conscientiously read labels stored in the over-the-counter cabinets. If you must do a great deal of close work at eye level or above, the problem can be solved by having the reading segment of the bi-focals placed at the top of the lens instead of at the bottom.

If you have had attacks of major muscle symptoms in the past, you should by all means avoid any job that involves heavy manual labour. If you have not had severe symptoms, but have not been used to doing a great deal of lifting or pushing, start slowly on a job requiring these activities. Don't jump in all at once and place a strain on your muscles that will require days or weeks of soreness until they recover. And get in condition for this kind of work by stretching the muscles of your back and legs, as in ortho-exercises 3, 4, 5, 6, 13 and 21 in Chapter IV.

If you are looking for a job, be sure to test yourself for any musculo-skeletal imbalance before you choose the kind of work you will do. If you have positive hip tension signs you should definitely not take a job requiring continual heavy lifting, back bending, pushing or pulling.

You and Your Muscles at Home. 'I wrenched my back mowing the lawn, doctor, and I can't even straighten up.'

'I don't know why, but whenever I load or unload the dishwasher I get this awful pain in my hip.'

'I was shovelling snow and my leg suddenly gave way right under me.'

'After I vacuum the house, I can't do another thing for hours – I'm just no good for the rest of the day.'

Not a week goes by when I don't hear one of these statements, or one just like them. Men are just as likely to come in with these complaints as women, for such activities as washing the car, raking leaves, putting up wallpaper or laying down floor coverings have proved as likely to cause painful symptoms as making beds, moving furniture, scrubbing floors and washing windows. Furthermore, it seems that the universal distaste for all of these necessary but tedious jobs causes us to attack them energetically in the hope of getting them over quickly – and our muscles are just not up to these sudden bursts of wear and tear, stress and strain.

It is virtually impossible for most of us to avoid tasks like these. However, those who have had serious muscle imbalance problems should make every effort not to do them. Try to get someone else, friend or employee, with a stronger back than yours to put up wallpaper, paint the walls, wax the floors and move the piano to the other side of the room.

A little ingenuity can go a long way towards relieving your aching back of an unnecessary burden. For instance, in making the bed, try a squatting position for placing and tucking in sheets. And it might be worth your while to have your local handyman or carpenter raise your bed a foot or more. This will not only save you bending over every time you want to smooth out a wrinkle, but it may provide the added dividends of hidden storage space under the bed, and a new centre-of-interest in your room that will make you the envy of any interior decorator. If the worst comes to the worst, of course, you can always leave the beds unmade!

Simply learning the correct way to hold normal household implements and appliances will reduce muscle strain considerably. When you use such implements as hoes, rakes, mops or vacuum cleaners, your body repeats the same motions over and over again for relatively long periods at a time. This causes tension and fatigue. Learn how to use your body in such a way that tension will be minimized. When you push and pull the tool back and forth in front of you, your body tends to lean forward, putting constant tension on the back and shoulder muscles. You will avoid a backache by standing sideways with your feet fairly wide apart, and using the implement in a left-to-right motion instead of forward and back.

When using a spade or snow shovel, the strain comes from lifting the heavy weight at the far end of the lever composed of your arm plus the shovel handle. By shortening the lever you can lighten the strain. Slide one hand as far down the shaft of the shovel as possible. Use this hand as a fulcrum and the other hand as the working force by pushing it down on the end of the handle. When you have a really heavy load on the shovel, you can bend one knee and brace the handle against your thigh as a fulcrum. You can further ease the strain by standing with your feet apart to widen the base of support and by bending at the knees instead of at the waist.

When doing such chores as ironing or washing dishes, the relative lack of movement while you stand in one position for a long time is what causes strain. Here, too, there are several ways to reduce or eliminate the muscle strain. First, make sure your working area is well lit. The job will go faster and your head will not be straining forward to see more clearly what you are doing. If possible, do chores from a sitting position. A kitchen stool is just the right height for most ironing boards (many of these are adjustable in height, too), and these perches can sometimes be pulled right up to the kitchen sink as well. Otherwise, keep one foot up on something while you work. Sometimes the floor of the cabinet under the sink is high enough to serve as a built-in footstool.

When you are doing household chores, wear garments that do not pull on you as you reach and stretch. Your

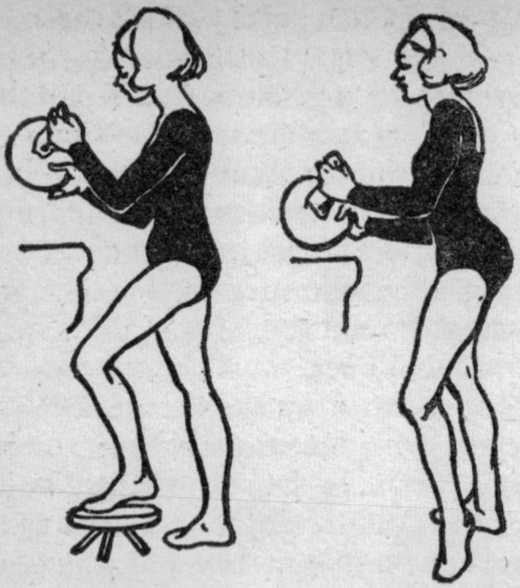

CORRECT INCORRECT

muscles have enough work to do without struggling against your clothes.

Everyone will benefit from sleeping on a firm mattress; and for those with backache problems, support at night is particularly important. Never buy, and try not to have to sleep on, a soft mattress of the kind that you 'sink into'.

CORRECT INCORRECT

If you are stuck even for one night with a hotel mattress that is too soft, you will know it the next morning in the stiffness and soreness of your back and shoulders. When this happens – which it does to all of us on some occasions – try to get another bed brought in. Frequently the 'rollaways' that hotels keep for the use of children are harder and better for your back than the regular bed mattress. In an emergency, try folding a bedspread or a blanket and placing it over the mattress. Locate an extra pillow and place it under your hips. (I don't know why, but hotels with flabby mattresses usually seem to have pillows that are hard as rocks!) And if you do a lot of travelling, for business or pleasure, you will probably find it worthwhile in terms of your health and comfort to invest in a folding bed board that can be placed under any mattress to give you firm support and a good night's sleep, with no achy after-effects the next morning.

For a bed board at home, order from the nearest timber yard a sheet of plywood three-quarters of an inch thick, cut to just one inch less than the dimensions of your mattress. Place it between the box spring and mattress, and sleep free of shoulder, neck and low back pain.

CORRECT

INCORRECT

If you have back trouble, learn how to live without a hollow in the lower part of your back. Sit with your behind tucked away to eradicate the hollow in your back. Sleep on your back with your knees propped up over a pillow, or on your side with one or both knees drawn up. Always squat, never bend over, when you have to pick something up. Never sit rigidly upright in a sway-backed position. Don't slump forward when you drive; move your seat close enough to the steering wheel for your knees to be bent all the time. Avoid wearing very high heels as much as possible.

Remember to keep on a maintenance programme of exercise so that, once your muscles have been strengthened, they never get a chance to relapse. And bear in mind this rule: *always* bend your knees; *never* bend your waist.

Muscles and the Military. At no other time do muscle problems show up as dramatically as when young men enter the armed forces. Not only do many men called for physical examinations fail to pass, but of those inducted, many men have to be discharged shortly afterward because they are physically unable to get through basic training. According to one report, more than 40,000 men inducted during a two-year period were later found to be physically unfit and had to be discharged, costing the government an unnecessary $40 million. Early recognition of musculo-skeletal imbalance during the early rapid growth spurts, and appropriate management, would do much to provide our military with physically fit candidates.

The backbone of the soldier is often the military's weakest point. Lower back pain is one of the most frequent complaints heard at sick call, and a leading cause of lengthy absence from military duty.

Another weak point is the soldier's feet. Foot problems show up as soon as basic training begins. Those men with good muscle function can go through basic training without problems. But those with muscle imbalance (thirty per-centers) often don't make it. They may be labelled as shirkers because they can't carry out basic training, but the truth is they simply can't march for long periods. March fractures are very, very common. At the end of nearly every day's march, in each troop, dozens of men are found with one or two fractured bones in their feet. Sometimes even a leg bone can get fractured from the strain of prolonged marching.

One captain reported that in his short experience there were almost 300 foot fractures just among men on assignment to basic training camp.

The severe pain that is a fracture symptom may occur immediately, but it also may not show up until a week or more later. A good bit of the problem could be prevented if adequate boots were issued to servicemen. During World War I, the Prussian army increased their marching efficiency dramatically with their well-designed (General Blücher) boots. The Prussian soldiers could march thirty to thirty-five miles a day carrying a seventy-pound pack over terrain on which the American troops couldn't even march eleven miles with a much lighter pack. In my opinion the major cause of march fractures can be found among the vulnerable thirty per-cent-ers.

Many new soldiers find that basic training brings on recurrence of painful knees or painful hips which they had when they were younger. Muscle-strengthening exercises will seem extra tedious after the hours of physical exertion that men in basic training must put in, but they will pay off in greater health and comfort, especially if corrective footwear were to be made available for the thirty per-cent-ers.

Muscle Balance and Sports. Mary, at the age of fifteen, was on her way to being the star athlete at her all-girls high school. She loved all sports and played on every team she had time for. Then she suddenly developed pain in her hips and couldn't skate, play basketball, run or swim.

Several months on an intensive exercise programme were no effort for her, and she understood how each of the exercises I assigned was working on her muscles. It wasn't long before her pain went away and she returned triumphantly to her school athletic activities. But she and her parents were concerned that the condition might return, so she kept up the exercises. She became so agile and well coordinated that when she finished high school a few years later, she tried out for the Rockettes, the dancers at Radio City Music Hall, and she was accepted. One day she invited me to watch the Rockettes rehearse. She said she had a surprise for me. Wondering what she had in mind, and always curious for a new experience, I accepted her invitation.

So there I was, rattling around all by myself in the empty theatre, watching fifty scantily clad, long-legged young ladies go through their warm-up session before the actual rehearsal began. I was absolutely astonished. The exercises they were doing were almost precisely the same as I teach my patients. So that was Mary's surprise for me! No wonder she was such a natural for the Rockettes.

Good muscle tone and balance has a direct relationship to athletic ability. Conversely, muscle *imbalance* hinders athletic ability and enjoyment in two ways. First, the person with poor muscle function is almost certain to be awkward and clumsy at sports, and to learn to 'count himself out of the game' at an early age – if his playmates don't do it for him, that is. And second, muscle imbalance is a leading cause of sports-connected injuries.

Sometimes, an injury occurs in a person whose muscle imbalance problem was previously hidden. I recall a resident I worked closely with at the hospital, who refused to believe me when I told him he had tibial torsion syndrome in both legs and was sure to have trouble at some time because of it. He answered that he had been a star athlete at the university and captain of the basketball team. Later he was inducted into the army – at the age of thirty-four, after nine years of medical training during which he had not participated in any sports, quite out of condition. I knew that he would not only suffer the usual rigours of basic training, but would also soon be out on the basketball court playing with younger men. It wasn't long before I received a telegram from him, saying 'YOU WERE RIGHT LETTER FOLLOWS.' It seemed that in his very first game his feet slammed to the floor as his body lunged forward and he ruptured both heel cords. Surgical repair of both heel cords was carried out. I sent him diagrammed exercises and a prescription for corrective footwear.

In a study reported by the American College of Radiology, basketball is cited as the most common cause of rupture of the Achilles tendon. But this kind of rupture of a muscle or tendon is common as well in tennis players, runners and all ball players. Rupture of this tendon is apt to occur during backward running while playing baseball, handball or skiing. Whenever the rupture occurs spontaneously and without undue exertion, it is almost invariably due to muscle imbalance in a thirty per-cent-er without appropriate footwear.

Pitching is similarly hazardous for the baseball player. The muscle strain that comes from repeatedly delivering high balls often causes permanent damage to the elbow. In a study of professional pitchers about half were found to have some shortening of the muscles and tendons so that their elbows could no longer be fully extended.

Even something as seemingly simple as a hoop can give trouble to a person with muscle imbalance. When the hula hoop was popular, so many adults and children developed ruptured muscles, pulled tendons, and dislocated knee caps that the problem was written up in several medical journals.

The 'twist' did the same thing. Suzy was my first twist patient. She had a dislocated knee cap and her leg had to be placed in a cast. Then two boys came in with torn cartilage that looked like the sort of thing they would get in a football game. We found that most of the kids who developed knee problems from doing the twist were those with muscle imbalance.

Even jogging can give problems. Fatigue fractures from jogging are being seen more and more by orthopaedic surgeons as jogging becomes increasingly popular. Fractures may take place in the forefoot, heel, ankle, and occasionally even in the leg or hip. The fractures are most likely to occur in overweight individuals or in those who have been physically inactive for many years but attempt to jog several miles on the first time out.

If you want to join the joggers, first take the tests in

Chapter III. If you have muscle imbalance, start on a series of corrective exercises *before* you start your jogging programme. Gradually, as you strengthen your leg muscles, you can start jogging. Be sure to wear corrective footwear. In this as in any other athletic activity, don't over-exert yourself. Always start slowly and for a brief time; gradually, over several weeks or even months, work up to a more rigorous programme. If at any time you feel uncomfortable or develop any related problem, cut back to an easier level and build up your pace again.

The problem of athletics-related injuries caused by muscle imbalance has been recognized and discussed by school and professional coaches as well as by physicians.

At a recent meeting of the American Medical Association the Jets physician, Dr Jim Nicholas, urged exercise programmes to compensate for muscle imbalance and help prevent the fifteen to twenty million sprains and dislocations that Americans suffer every year in sports and school activities.

In another talk at the same conference, Dr Joseph S. Torg, a physician at Temple University School of Medicine, also warned about sports injuries. 'The incidence and severity of knee injuries among scholastic, collegiate and professional football players has become quite significant,' he noted. 'Those involved in the care of these athletes are well aware of the magnitude of the problem.'

In its report, the President's Council on Youth Fitness noted that seventy per cent of American schools have inadequate physical education programmes, and that the average youngster spends less than one per cent of his time in physical activities and ten per cent watching television. Even more discouraging is the realization that what little exercise the child is doing is too often the wrong kind. The President's Council and the American Medical Association have both called for screening of all school children to determine those in need of special exercise and health care; but so far almost nothing has been done. In the meantime, the tests and exercises in this book will help you to screen your own child and determine what his exercise programme should be.

Several football coaches are already screening their players. As a result of the routine muscle tests done on the Jets, their practice sessions now include conditioning and corrective exercises. Based on the test results, the men are divided into three or four groups to work together on related muscle problems.

In football and baseball a major cause of injury is the strain put on the knee joint when the player has his foot fixed firmly on the ground and twists his body or is tackled suddenly by another player. Since his shoe cleats are riveting the athlete's foot to the ground, a shoe with fewer and shorter cleats will decrease, but not eliminate, this hazard. Tests of a moulded-sole soccer type of shoe with short cleats showed that knee injuries to players who wore it occurred less than half as frequently as to those wearing standard cleated football shoes. Furthermore, those injuries that did occur were much less severe with this type of shoe. A shoe with quarter-inch cleats instead of three-quarter-inch cleats can thus make a big difference in the injury rate. Such shoes have already been adopted by the high school football leagues in Philadelphia and several other cities.

Another new type of footgear for ballplayers is the swivel shoe. The heel is cleatless, but there are movable or swivelling cleats on the front of the sole. This relieves the knee and ankle of the strain resulting from twisting action, allowing the foot to twist with the body.

Those who are not professional athletes but enjoy participating in active sports should choose their sports carefully if they have had evidence of a muscle-related problem. People who say they have weak ankles and can't ice skate well or ski well are probably right, and are sensible to recognize the problem. If skiing or ice skating is particularly important to you, you can help overcome such a problem with proper shoes. Otherwise, just relax and enjoy some other sport. Many people whose muscle imbalance problems hinder more vigorous activities choose swimming, golfing or boating.

Those who have tibial torsion syndrome can still enjoy basketball, tennis and running, provided they wear adequate supportive shoes with a half-inch cup doughnut lift in the heel to provide balance and compensate for the twist of their knee and ankle. (Your shoe shop can insert this for you.)

Swimming is excellent for everybody. The buoyancy of the water allows muscle action free of the gravitational pull that otherwise affects all movements. Many people with muscle imbalance have been so poor at most other sports that they are too discouraged to learn how to swim. On the other hand, many of these people are delighted to find that swimming is the one sport at which they can excel. In addition, most swimming strokes are excellent supplements to an ortho-exercise programme. Simply holding on to the side of the pool and kicking is one of the best hip exercises ever devised. However, the breast stroke overdevelops the wrong muscles, and should be avoided. Crawl or side stroke are both good, but back stroke is best. Jumping in or diving should be avoided by those with severe muscle imbalance.

The Importance of Shoes. Shoes are extremely important to proper posture and good muscle functioning. The wrong shoes can be a major cause of muscle aches and pains. Properly prescribed shoes in conjunction with exercises can do wonders for eliminating aches and pains.

People who are especially vulnerable to muscle and joint problems should never wear tennis shoes for sports, and should not go barefoot any more than necessary. Firm support is essential. Heel pads in shoes to decrease muscle strain are very helpful, but the pads must be prescribed precisely to fit each person's problem.

Many people with tired or aching feet, noticing that the heels of their shoes wear down on only one side, have bought special wedges or pads for their shoes. Then they may find that their feet hurt less, but their backs hurt more. This may even happen when a physician prescribes corrective footwear, if both the doctor and the patient don't watch for side effects or complications.

When I first started my practice, I used to treat flat feet, bow-legs, or knock-knees using the same procedures everybody else used – a $\frac{3}{16}$-inch inner heel wedge and a cookie insert, a Thomas heel or other corrective shoe. That's what we had been taught was the traditional and proper treatment. For some patients it worked. But in others, the symptoms became much worse because of the correction.

I can remember when the implications of this first occurred to me. A mother had been bringing her two-year-old to the office for his foot problems – unsteadiness and stumbling when he was trying to walk, and flat feet. I prescribed the usual $\frac{3}{16}$-inch inner heel wedge and a straight last support shoe.

After about nine months, the mother came in very disturbed. She said 'I want to know exactly what you are treating in my child. Just look at his knees.'

To my surprise, the child was knock-kneed. I looked back on my chart and there was no mention of knock-knees. I check for this as a matter of course and make a note of its presence. But there was no mention of it on either the first examination or the following ones. I did not want to admit my puzzlement. 'Continue the treatment for flat feet,' I said, 'and return in four months and we'll take care of the knock-knee situation then.'

But after they left the office I sat and tried to work the puzzle out. The inner heel wedge that I had prescribed for the child was the standard treatment for flat feet. And if I hadn't prescribed it, the paediatrician would have, or the general practitioner, or a podiatrist or a shoe salesman. And, if I had noticed the knock-knees earlier, what treatment would I have prescribed? Precisely the same – the inner heel wedge, perhaps a pad under the arch, and an inner sole wedge.

About ten days later I saw another mother and child of about the same age. This little fellow's foot problems included constant falling, and turning his toes in. His shoes wore down completely on the outer side of the heel in only two or three weeks. On his last visit I had prescribed a wedge. I was particularly eager to examine him. And sure enough, he was beginning to develop knock-knees.

I started watching for it in other patients, carefully checking each child on every visit. After a while I worked out that the knock-knees were a compensation for the additional strain placed on these young legs and muscles by the heel wedge.

Another piece of the puzzle fell into place when my own shoes wore down at the heels and I took them to the shoemaker for new heels. After he had mended them, I developed a pain on the inner side of the knee which continued for several days. I connected the knee pain with the new heels on my shoes, and I looked at the shoes more carefully than when I had taken them from the shoemaker. He had added one eighth of an inch to the inner side of the heel! Removing the heel wedge gave me immediate relief from the knee pain.

Part of the problem is the hard floors and pavements of our modern, urban environment. When I was in the Middle East and in Africa I saw natives who had been born in the country and walked barefoot on the soil for all their lives. They had well-balanced, thick-soled feet. Not one complained of aching feet or pains in his legs. However, those who had moved to the cities suffered the same aches and pains as so many of the rest of us. The human musculoskeletal system was not designed to take the wear and tear brought on by the constant jarring of walking on concrete.

What kind of shoe should you have if you are among the thirty per cent of people with muscle problems? First, you must make sure that the instep is sufficiently wide and deep and roomy for the widest part of your foot. Your shoes should have a long counter – that is, the side of the shoe should cover the instep area for support. A straight last will curb turning the toes in or out. The bottom of the shoe should be leather, with a Thomas heel – that is, a rubber heel extended forward on the inside for greater support. Your doctor may recommend additional features, such as a steel bar to support the instep, or a heel pad. Shoes should be neither too flat nor too high.

It isn't how much money you spend for footwear, it's getting the right kind. Some medium-priced shoes are better than some expensive ones. The important thing is that they give firm support and are properly fitted.

It used to be extremely difficult to find a shoe of this type unless you were willing to wear the clumsy-looking, little-old-lady shoes shown in orthopaedic shoe shops. But then the fashion changed to broad, medium-heel styles for women, which are good for the feet as well as good-looking. Of course, fashions are changing again. But I doubt if the women who have got used to the broader type of shoe will willingly return to the cramping stilts they used to wear! Men's shoes are generally easier on the feet since they are more supportive, but men too should look for the same characteristics of long counter, wide enough toe and instep, and leather soles.

Obesity. Everyone must know by now about the excess strain that obesity places on the heart and circulatory system. It should not be surprising, therefore, to learn that obesity places an excess strain on the muscles and joints as well. If you have a muscle imbalance problem and are overweight in addition, you should get rid of those extra pounds.

Getting rid of excess weight is one way to help get rid of low back pain. One report at a recent meeting of an orthopaedic society listed three important treatments to attack low back pain: exercising, improving posture and maintaining proper weight. More than ninety-five per cent of 749 patients treated with that combination over a fifteen-month span were relieved of pain. Sometimes, a loss of only ten pounds can bring about relief.

The ortho-exercises described in this book are *not* weight-reduction exercises. Although they will use up calories, this is incidental to their main purpose: to firm up muscles and increase muscle and joint flexibility.

In fact, there *are* no weight-reduction exercises. The *only* way to lose weight is to diet. Make sure you still get a balance of vitamins and minerals, because these are essential for keeping the muscles and other body tissues healthy.

CHAPTER TEN

Prognosis: Positive

The aches and pains of the muscles and bones we are talking about in this book are physiological phenomena. But emotions can play some role in the problem and we must consider them in a discussion of the overall treatment. When we are emotionally tense and upset we tense our muscles. When we are anxious or angry, we may tense the muscles of the neck and shoulders. All of us have at some time had a stiff neck or a headache that we knew very well was due to the emotional tension we were under that day.

Not only can the emotions have an effect on body pain, but in turn the pain and disability can have an effect on emotions. No one who feels the dull grind of pain for most of his waking hours can be consistently cheerful and open to life. The personality is bound to be affected. Some people deny or underplay their pain and difficulties because they do not wish to admit to any body faults. Many have spent years feeling clumsy, inadequate and somehow different from other people – all because of muscle imbalance. These people too easily become insecure and self-conscious, suffering needlessly throughout their lives.

Often people with chronic backache or headaches are accused of having imaginary pains. They are told they are 'just nervous' or 'it's all in your mind'. They may themselves begin to believe that 'it's all in their mind', and may develop a real neurosis as a result.

If someone whose vague aches and pains have not been diagnosed takes the muscle tests in Chapter III, he will have proof that there is indeed a real physical ailment underlying his symptoms. He will finally be able to get rid of that nagging suggestion that perhaps he is a little neurotic. It's very satisfying when you can say 'I *knew* there had to be a cause – I *knew* it wasn't in my mind and now I've solved the problem.'

Even more satisfying, however, will be the relief from pain that he will feel when he is on an ortho-exercise programme. I am not exaggerating when I say that I have seen personalities change from 'grouchy' to cheerful in a matter of months. I have even seen faces change – the tight, intense, inverted expression of someone in pain becomes the open, expressive face of someone looking forward to life, as the patient experiences first hours and then days free from pain. In many cases, these changes are so complete as to be truly astounding.

What kind of results can you expect from orthotherapy?

I have treated thousands of patients with muscle problems in my years of practise as an orthopaedic surgeon. And when patients follow advice and keep up with the programme, the changes have been astounding.

The results of orthotherapy used to correct improperly working muscles have been reported in various scientific journals and at meetings of orthopaedic specialists. It has been observed that the degree of correction is directly and specifically related to the age of the patient, how diligently he performs his exercises, whether he has been cooperative in carrying out other suggestions concerning treatment, and how severe the problem was to begin with.

But more impressive than these statistics are case histories that show how an individualized exercise programme, combined with other recommendations as necessary, can make an overwhelming difference in a person's life.

Larry Sherman is typical of a large group of patients. When I first saw him, he complained of a steady, dull ache in his lower back. He had difficulty in bending over, in walking, and sometimes felt pain radiating down from his hips to the back of his thigh. At first the backache could be relieved by rest, but soon that didn't help either. He thought he was stiff-jointed, and that he could loosen up the muscles with callisthenics, but the pain steadily became worse. The stiffness and soreness was usually most intense when he got up in the morning, or after any period of inactivity. He had taken aspirin to get some relief; but after several months it no longer seemed to have any effect.

We used a combination of treatments: an exercise programme to retrain muscles; rest in bed and aspirin (I try to avoid stronger medication if possible) when severe attacks came on; and applications of heat to the affected area. In severe cases, a week or more of continuous traction in the hospital is sometimes necessary. But this, too, I try to avoid. And my patients usually find, as did Larry, that the rest and the heat are enough to reduce the severity of his acute attacks, and that over a period of time the exercises do the rest.

Within five months Larry was no longer having periods of crippling back pain. He could straighten up and walk easily and he no longer lived in fear of sudden stabs of agonizing pain. He knew that, using the exercises as he had been shown, he could stretch his muscles out of their cramping whenever he needed to.

Another typical combination of symptoms was seen in the case of Alice Blake. She complained of pains in her legs, and very low in the tail bone area of her back, which was becoming increasingly stiff. She said the trouble had started when she tried to lift a heavy grocery carton. But when I took a detailed medical history, it emerged that she had minor symptoms of a weak back all her life. Her problem, however, had remained relatively minor, and an exercise schedule was all Miss Blake needed to relieve the pain and stiffness.

One of my most gratifying cases was that of Sarah Barnes, a sweet-faced woman in her seventies. One day while she was walking her foot suddenly caught on the grid of a grating in the pavement. The next thing she knew she was lying in the middle of the pavement with a broken hip.

Sarah thought she had broken her hip because of the fall, but actually she fell because her hip broke. When her foot had suddenly caught on the grating, it unexpectedly stopped her in the middle of a step. However, the momentum of her body weight caused her to continue forward, her foot immobilized while her body was still moving, producing a twisting, whiplash action on the hip bone and causing the bone to fracture at a weak spot. This is a common occurrence, particularly in older people, and there are thousands who believe they have fallen and broken a hip, while they actually broke the hip and *then* fell.

Sarah was taken to a hospital where X-rays showed a fractured hip. The hip was set and a surgical nail was used to hold the broken bone in place. But Sarah's set bone would never have healed completely without the exercises we had her do while it was mending. If pain is felt in the hip four to six months after a fracture appears to have healed, it means that the bone has not really healed.

Because this is such a common problem, we routinely have our patients begin exercises two or three days after the setting of a fracture or after surgery. The surgical nail is a boon for orthotherapy of fracture patients, because it permits movement of the limb without disturbing the set bone. We try to get elderly patients out of bed as early as possible, walking with crutches as soon as X-rays indicate that it is safe. Usually full weight-bearing is delayed for five to six months, and in the majority of cases there is sufficient union of the bone by that time to permit walking without support.

So, when Sarah Barnes asked me whether her hip would ever be 'as good as new' again, I was able to explain all this to her, and promise good results. She was eager to aid in her own return to normalcy by doing exercises as I prescribed them for her.

An eight-year-old-boy was brought to me by his parents, Mr and Mrs W. The boy was one of those non-athletic kids who loves to watch sports, knows all the rules and the names of all the players, but just seems to be an awkward bunch of bones on the field. He was all thumbs and had three left feet when he tried to play any game. He had had difficulty in learning to walk, his parents said, always tripping over his own feet, falling and stumbling, and lately he was having problems in school because he couldn't sit still for more than a few minutes at a time. When he did his homework, he constantly fidgeted about because his muscles were tense. And he was always tired and irritable.

After taking his medical history, I gave him a thorough examination. While talking to his parents I learned that Mrs W. had a complaint, too – frequent pain in her neck and shoulders. When I examined her I found muscle spasm in some places along her neck and shoulders, and exquisite tenderness when I pressed on the muscle.

I started both the boy and his mother on exercise programmes – different ones for each. After five months of exercising regularly, the boy was quite different. He could finally hold his own with the other kids in games and sports, and the improvement delighted him so that he was glad to continue exercising. And Mrs W., too, found her muscle spasms relieved.

It's curious how often one case leads to another and another. Mrs W. told a neighbour of hers about the improvement in herself and her son. The neighbour, Mrs Parker, brought her daughter Mary to my office. Mary was the kind of child that teachers dread to see – she simply couldn't sit still for five minutes. She was constantly fidgeting, stretching, jumping up and down in her seat. Naturally, she couldn't concentrate on her schoolwork. So she absorbed less and less of what was being taught, fell further and further behind in class – and continued fidgeting. When she was doing her homework, she'd sit for a few minutes and try to concentrate, but she would soon be out of her chair and wandering around. Her mother was constantly yelling at her to stop fidgeting, and sit down and study. But Mary's trouble was in her muscles, not in her head. She was physically unable to sit properly or comfortably. A personalized exercise programme to retrain her muscles solved her problems, too.

Another patient was referred to me by her gynaecologist

because, doing a routine visit, she mentioned a problem that she had suffered with for three years. Helen S. told her doctor that she experienced great pain in her hips and muscle spasms in her legs during intercourse. The pain was quite severe. And, she continued, it got worse by the month. Now, after three years of marriage, she and her husband were both nervous and exhausted and possibly on the verge of a mental breakdown and a marital breakup.

Examining her internally, the gynaecologist was able to rule out any problem originating from her sexual organs. Fortunately, he had become interested in orthopaedics. He put her through some simple tests and found the muscles controlling her hip and thigh movements were very rigid. And he had noticed her grimace of discomfort when he had placed her in position for her examination. She had been unable to spread her legs wide enough for comfortable intercourse to be possible.

She was delighted to start on an exercise programme; she went at it conscientiously and enthusiastically, and very shortly had the great pleasure of knowing that she had resolved a problem that had only a few weeks earlier threatened to wreck her mental stability and marriage.

Orthotherapy and Serious Illness. Orthotherapy exercises can be important also in retraining and restrengthening muscles in many diseases and disorders that affect the muscles and bones.

Exercise is essential for arthritis patients, for example, to make the joints function more efficiently. Exercise is particularly helpful following a 'flare' – a severe inflammation of the affected joint. Tissues loosened during a flare are often tightened up again in the wrong position. Exercises prescribed for rhythmic flexing and relaxing of affected muscles will gently pull the still tender joints into their normal alignment. The exercise not only prevents the joint from 'freezing' into unsightly and useless positions, but also strengthens connective tissue weakened by the inflammation and inactivity. There should be no exercise *during* an arthritis flare, however. Corrective footwear is always helpful to the thirty per-cent-er.

Exercise is also helpful in Parkinson's disease (shaking palsy). Parkinson patients frequently have crippling contractures of tendons and muscles, especially of the feet, legs, arms and hand. Whether the patient has surgery or uses the new and very effective drug L-Dopa or not, he should do exercises once or twice a day as soon as his doctor permits. This will bring back his range of motion and stretch and relax the muscles that have been in contracture.

Muscles are also contracted and tight in many patients who have cerebral palsy. In fact, I call cerebral palsy the most perfect form of muscle imbalance. In these cases, too, exercises can help loosen the contracted muscles and increase the range of normal motion.

Other Medical Treatments. Drugs, manipulation or massage, brace or cast, surgery – only a physician will be able to decide if any of these treatments are needed in a given case.

Several drugs that are very beneficial in treating muscle and joint pains have come on the market recently. Muscle relaxants have relieved hip pain many times when nothing else has helped. Hydrocortisone or procaine injections often help relieve the pain of osteo-arthritis during an acute attack.

All medications, of course, should only be used under the direction of a physician.

Manipulation must be done by your physician, as only he knows what kind you need.

The purpose of manipulation is to relax and stretch muscles and consequently reduce pain from tension. Manipulation should never be undertaken without first performing the dynamic tests. For shoulder and neck pain, for example, fatigued neck muscles can be stretched by manipulation. With the patient lying on his affected side, the physician presses down and outward on the shoulder blades with the heel of his hand. If the pressure is sustained for five or six minutes, the patient will experience relief. Manipulation alone will not cure muscle imbalance, but with supportive ortho-exercise therapy, it can be very helpful.

Sometimes casts or braces are necessary in addition to ortho-exercises. They allow the muscles to relax by supporting the affected joint. The results are particularly good in young children who still have a great deal of elasticity in their bone and muscle systems. When casts or braces are used, they may have to be changed a number of times, as the child's body grows or as the bones and muscles begin to respond and change. A rubber heel is usually incorporated into a leg cast so that the patient may be up and around.

See Your Doctor. You need not suffer daily aches and pains, and ortho-exercises can do more towards eliminating them than you would ever have believed possible. But it is still best to see a physician also. He will examine you to be sure that some underlying disease is not the cause of your complaints. If his examination confirms the need for the exercises, he can give you an individualized schedule of exercises and demonstrate how to do them. And he can prescribe drugs or other medical aids to lessen discomfort or speed up treatment.

You can go to an orthopaedic surgeon, to a specialist in physical medicine (now called a physiatrist), or to your regular family doctor. If your regular doctor cannot help you, do not hesitate to ask him to refer you to a specialist in this field. No one physician can be equally expert about every facet of the human body, and a competent physician has no qualms about referring a puzzling case to someone else who might be able to help. If you do not have a regular physician and you do not know where to turn, you can find a list of General Practitioners at your local Post Office; if necessary a GP will refer you to a specialist or to hospital.

Before seeing your doctor, jot down in advance the symptoms or facts about your medical history that you want to be sure to mention. If you have specific questions, make a note of them too. Also, if your doctor doesn't give you a memo of his recommendations, take your own notes while he talks to you. Many a patient has walked out of a physician's office and realized five minutes later that he can't remember all the instructions he was given. If necessary, call back later to get a point clarified.

Don't worry about how much of your doctor's time you are taking; tell him all of the details, even those that are embarrassing or may seem only remotely connected with your condition. It is astonishing how often a symptom mentioned casually can be the key to unravelling the mystery of a disorder. And if you have any specific ideas as to what might be causing your particular disorder, do voice them. Whether by hunch or insight or just plain luck, it is amazing how often patients do have the answers to their own problems.

For the Future. I would like nothing more than to be able to improve the condition of the musculo-skeletal systems of the entire population! In fact, I have a five-point plan that would go far towards accomplishing this:

1. Examination of *all* children at the time they enter school, to detect muscle imbalance and unfitness. (Just as children are routinely examined for such things as heart, vision and hearing defects, they must be examined for muscle defects.)

2. Periodic re-examination at crucial stages of development. (Minimally, muscle fitness should be checked at both the beginning and near the end of adolescence; for practical purposes, examinations for the youngest and oldest children in secondary schools should not be difficult to achieve.)

3. Complete revision of the standard physical education programmes now in use in the schools, with emphasis on musculo-skeletal imbalance detection and muscle-stretching callisthenics for coordination, not endurance. (This, of course, would require the additional training of physical education personnel; the introduction of new courses in teacher training colleges; and perhaps the development of a corps of para-professional personnel with specialized skill in this area.)

4. Exercise 'centres' or clubs or classes *in every community*, accessible to everyone, manned by a trained orthotherapist. (Everyone would attend group sessions geared to their needs; the orthotherapist would work with local physicians and train patients in exercises prescribed.)

5. In fact we should have a public health physical monitoring system which would include periodic dynamic testing for every citizen from birth to death.

My entire philosophy would be meaningless and my years of teaching and practice wasted if I did not try to explain to each of my patients how muscle imbalance affects them and why ortho-exercises are so important. I hope that with this book, as well as through my teaching and my practice, I can reach enough people to ensure that true physical fitness will one day be a reality.

Main Muscles Considered in Orthotherapy

Adductors. Technically, any muscle which pulls a bone towards the mid-line of the body is called an adductor. The adductor referred to in this book goes from the pubic bone to the mid-shaft of the thigh bone and is called the adductor longus.

Biceps. A prominent muscle on the front side of the upper arm which extends from the shoulder and wing bone to the front of the forearm bone.

Gastroc-Soleus. The gastrocnemius is the large muscle which gives the round form to the calf. It goes from the back side of the lower thigh bone to the heel bone, forming the heel cord. The soleus is an associate of the gastrocnemius and is located just in front of it.

Gluteus Maximus. A very large, fleshy muscle at the back of the hips (buttocks) which extends from the outer part of the thigh bone to the hip bone.

Hamstrings. Muscle strips readily seen on the back of the thigh just above the knee joint which go from the hip bone to the upper end of the leg bone.

Iliopsoas. Actually a composite of two muscles, the psoas major and the iliacus. The psoas major is in the abdominal cavity behind the internal organs and goes from the side of the last dorsal (chest level) vertebrae and all the lumbar vertebrae to the inner side of the thigh bone. The iliacus is a flat, triangular muscle which extends from the inner border of the hip bone to the inner side of the thigh bone.

Quadratus Lumborum. The four-sided muscle of the loins which forms a flat sheet on each side of the spinal column and goes from the lower border of the last rib to the crest of the hip-bone.

Quadriceps. The quadriceps is composed of the rectus femoris, the vastus lateralis and the vastas medialis. The rectus femoris goes straight down the front of the thigh from the hip joint to the knee cap. The vastus lateralis goes from half-way down the outer side of the thigh to the knee cap. The vastus medialis goes from the upper end of the inner side of the thigh bone to the inner side of the knee cap.

Triceps. The triceps is on the back side of the upper arm. It goes from the wing bone and two places on the upper shaft of the arm to the elbow.

Index